# Thermochemistry

# Thermochemistry

**Rose Marie O. Mendoza**

www.arclerpress.com

# Thermochemistry

*Rose Marie O. Mendoza*

**Arcler Press**

224 Shoreacres Road
Burlington, ON L7L 2H2
Canada
www.arclerpress.com
Email: orders@arclereducation.com

**© 2021 Arcler Press**

ISBN: 978-1-77407-761-0 (Hardcover)

Arcler Press publishes wide variety of books and eBooks. For more information about Arcler Press and its products, visit our website at www.arclerpress.com

# ABOUT THE AUTHOR

**Dr. Rose** obtained her PhD in Chemical Engineering from the University of the Philippines-Diliman in 2013, Her Masters in Chemical Engineering and BS Chemical Engineering degree from Adamson University. She is also a Professor in the Graduate School Department under the Master of Engineering Program at Adamson University since 2006, and is a Visiting Research Fellow at the Department of Environmental Engineering and Science, and the Department of Environmental Resource Management in Chia Nan University of Pharmacy and Science, Tainan Taiwan since 2010. She obtained her Post Doctorate Degree in Green Power: Hydrogen Generation and Fuel Cell Development from the University of California Merced, USA (in collaboration with the Energy Storage and Conversion Materials Laboratory of the University of the Philippines, Diliman) and is presently also a collaborating researcher at the Kindai University, Nara Japan, focusing on applications of electric and electrostatic field.

Dr. Mendoza has several international and local researches covering Fuel cells and solid oxide electrolysis cells (SOECs), green energy and green technology, biomedical and natural products, fuels and fuel development, at water and wastewater treatment, recovery and remediation. She also published several international textbooks on water management, introductory chemical engineering and chemistry. Presently, Dr. Mendoza is the Chief Science Officer of Mark-Energy Revolution Corporation, Philippines.

# TABLE OF CONTENTS

# GLOSSARY

---

A

Active electrode – electrode that participates in the oxidation-reduction reaction of an electrochemical cell; the mass of an active electrode changes during the oxidation-reduction reaction.

Actuarial – relating to actuaries or their work of compiling and analyzing statistics to calculate insurance risks and premiums.

Anode – electrode in an electrochemical cell at which oxidation occurs; information about the anode is recorded on the left side of the salt bridge in cell notation.

B

Basis set – the set of mathematical functions (basis functions) used to describe the molecular orbitals.

Biomolecule – molecules and ions present in organisms that are essential to one or more typically biological processes.

Boiling point – temperature at which the vapor pressure of a liquid equals the pressure of the gas above it.

Bomb calorimeter – device designed to measure the energy change for processes occurring under conditions of constant volume; commonly used for reactions involving solid and gaseous reactants or products.

C

Calorimeter – device used to measure the amount of heat absorbed or released in a chemical or physical process.

Cathode – electrode in an electrochemical cell at which reduction occurs; information about the cathode is recorded on the right side of the salt bridge in cell notation.

Cell notation – shorthand way to represent the reactions in an electrochemical cell.

Cell potential – difference in electrical potential that arises when dissimilar metals are connected; the driving force for the flow of charge (current) in oxidation-reduction reactions.

Clausius-Clapeyron equation – mathematical relationship between the temperature, vapor pressure, and enthalpy of vaporization for a substance.

Collision – A collision is the event in which two or more bodies exert forces on each other in about a relatively short time.

Colloquially – in the language of ordinary or familiar conversation; informally.

Compilations – the action or process of producing something, especially a list or book, by assembling information collected from other sources.

Compression – The action of compressing or being compressed.

Condensation – change from a gaseous to a liquid state.

Conducive – making a certain situation or outcome likely or possible.


D

Density – The density, of a substance is its mass per unit volume.

Deposition – change from a gaseous state directly to a solid state.

Diffusion – Diffusion is the net movement of anything from a region of higher concentration to a region of lower concentration.

Dihalogen – containing two atoms of a halogen element.

Disinfectant – a chemical liquid that destroys bacteria.

Dissociation – the action of disconnecting or separating.

Divalent – having a valency of two.

Dynamic equilibrium – state of a system in which reciprocal processes are occurring at equal rates.


E

Elastic – able to resume its normal shape spontaneously after being stretched or compressed.

Electrolysis – chemical decomposition produced by passing an electric current through a liquid or solution containing ions.

Electronegativity – having a tendency to attract electrons.

Endothermic – accompanied by or requiring the absorption of heat.

Energy – the capacity of doing work or supplying heat.

Enthalpy – a thermodynamic quantity equivalent to the total heat content of a system.

Entropy – the degree of disorder or randomness in the system.

Exothermic – accompanied by release of energy through light or heat.

Exothermic – (of a reaction or process) accompanied by the release of heat.


F

Freezing point – temperature at which the solid and liquid phases of a substance are in equilibrium; see also melting point.

G

Galvanic cell – electrochemical cell that involves a spontaneous oxidation-reduction reaction; electrochemical cells with positive cell potentials; also called a voltaic cell.

Germination – the development of a plant from a seed or spore after a period of dormancy.

H

Hydrolysis – the chemical breakdown of a compound due to reaction with water.

Hyperbolic – relating to a hyperbola.

I

Inert electrode – electrode that allows current to flow, but that does not otherwise participate in the oxidation-reduction reaction in an electrochemical cell.

Insulated – protected by interposing material that prevents the loss of heat or the intrusion of sound.

Irreversible – not able to be undone or altered.

Isolated – far away from other places, buildings, or people; remote.

Isotopes – variants of a particular chemical element which differ in neutron number, and consequently in nucleon number.

K

Kinetic Energy – the kinetic energy of an object is the energy that it possesses due to its motion.

L

Law of Conservation of Energy – in a chemical of physical process, energy is neither created nor destroyed.

Ligand – an ion or molecule that binds to a central metal atom to form a complex.

Lignocellulosic – plant dry matter.

M

Macroscopic – visible to the naked eye.

Manometer – an instrument for measuring the pressure acting on a column of fluid, consisting of a U-shaped tube of liquid in which a difference in the pressures acting in the two arms of the tube causes the liquid to reach different heights in the two arms.

Meagre – (of something provided or available) lacking in quantity or quality.

Melting point – temperature at which the solid and liquid phases of a substance are in equilibrium; see also freezing point.

Metabolism – the chemical processes that occur within a living organism in order to maintain life.

Molar heat – the amount of heat needed to raise the temperature of 1 mole of a substance by 1 Kelvin.

Molecules – A molecule is an electrically neutral group of two or more atoms held together by chemical bonds. Molecules are distinguished from ions by their lack of electrical charge.

O

Obscures – keep from being seen; conceal.

Oxidation – the process or result of oxidizing or being oxidized.

P

Plunged – jump or dive quickly and energetically.

Potential – having or showing the capacity to develop into something.

Pyrolysis – the thermal decomposition of materials at elevated temperatures in an inert atmosphere.

R

Reactants – a substance that takes part in and undergoes change during a reaction.

Reformation – the action or process of reforming an institution or practice.

Re-oxidized – to remove electrons from an atom or molecule.

S

Solidification – a phase change of matter that results in the production of a solid.

Spectrometrist – person who study about spectrometry.

Spectroscopy – the study of the interaction between matter and electromagnetic radiation.

Spontaneous – performed or occurring as a result of a sudden impulse or inclination and without premeditation or external stimulus.

Stoichiometry – a section of chemistry that involves using relationships between reactants and/or products in a chemical reaction to determine desired quantitative data.

Sublimation – change from solid state directly to gaseous state.

Submerged – cause (something) to be under water.

Surpass – exceed; be greater than.

Surroundings – All matter other than the system being studied.

System-portion of matter undergoing a chemical or physical change being studied.

T

Thermochemical Equation – An equation that includes the heat change.

Thermochemistry – Concerned with the changes that occur during chemical reactions.

Thermodynamicists – Person who does study of thermodynamics.

Transition – The process or a period of changing from one state or condition to another.

Transmitted – Passed on from one person or place to another.

V

Vapor pressure – (also, equilibrium vapor pressure) pressure exerted by a vapor in equilibrium with a solid or a liquid at a given temperature.

Vaporization – An element or compound is a phase transition from the liquid phase to vapor.

Variations – A change or slight difference in condition, amount, or level, typically within certain limits.

Vigilant – Keeping careful watch for possible danger or difficulties.

Volatile – Liable to change rapidly and unpredictably.

Voltaic cell – Another name for a galvanic cell.

# LIST OF FIGURES

# LIST OF TABLES

# LIST OF ABBREVIATIONS

| | |
|---|---|
| AgNO3 | Silver Nitrate |
| aq | Aqueous |
| B3LYP | Becke, 3-parameter, Lee–Yang–Parr |
| CCD | Coupled-Cluster, Doubles |
| CCSD | Coupled-Cluster, Singles, And Doubles |
| CCSDT | Coupled-Cluster, Singles, and Doubles and Triples |
| CI | Configuration Interaction |
| Cl | Chloride |
| CNG | Compressed Natural Gas |
| CO2 | Carbon Dioxide |
| DFT | Density Functional Theory |
| EA | Electron Affinity |
| GB | Gas Basicity |
| GHG | Greenhouse Gas |
| GIANT | Gas-Phase Ion and Neutral Thermochemistry |
| H2– | Water |
| HD– | Hydrodeoxygenation |
| HF | Hartree-Fock |
| Hfus | Heat Of Fusion |
| Hvap | Heat Of Vaporization |
| IE | Ionization Energy |
| IMF | Intermolecular Force |
| ITC | Isothermal Titration Calorimetry |
| K2Cr2O7 | Potassium Dichromate |
| KE | Kinetic Energy |
| kJ | Kilojoules |
| KMT | Kinetic Molecular Theory |

| | |
|---|---|
| kPa | Kilopascals |
| MM | Molecular Mechanics |
| M– | Molecular Orbital |
| Na | Sodium |
| NaCl | Sodium Chloride |
| NaNO3 | Sodium Nitrate |
| P | Pressure |
| PA | Proto Affinity |
| PCI-X | Parametrized Correlation Method, Containing A Single Adjustable Parameter X |
| QMC | Quantum Monte Carlo |
| SCF | Self-Consistent-Field |
| TES | Thermal Energy Storage |
| V | Volume |
| VB | Valence Bond |
| VSEPR | Valence Shell Electron Pair Repulsion |
| ZSM-5 | Zeolite Soconymobil-5 |
| $\Delta H_{con}$ | Enthalpy of Condensation |
| $\Delta H_{dep}$ | Enthalpy of Deposition |
| $\Delta H_f$ | Enthalpy of Formation |
| $\Delta H_{frz}$ | Enthalpy of Freezing |
| $\Delta H_{fus}$ | Enthalpy of Fusion |
| $\Delta H_{sub}$ | Enthalpy of Sublimation |
| $\Delta H_{vap}$ | Enthalpy of Vaporization |
| $\Delta T$ | Temperature Change |

# PREFACE

Thermochemistry is the field of science that combines the principles of thermodynamics and chemistry, both in theory and application. This combination of the two fields helps the society provide innovations that can help in finding solutions to several of the most challenging problems in the world.

Thermochemistry is closely associated with thermodynamics and is based mostly on the principles that are defined in conventional thermodynamics. It focuses on the utilization of heat in the various chemical processes and how the generation and consumption of heat can be calculated so that the scientists and researcher can get an exact idea of what is happening in those processes.

This book aims at making the readers aware about the subject of thermochemistry and explains the manner in which various principles of thermodynamics can be used to calculate several physical quantities, which can further be utilized in certain projects and technologies for further development.

This book informs the readers on the historical development of science the way in which thermochemistry has evolved by discussing some introductory and basic concepts in the field as discussed in the first chapter.

The second chapter takes the readers through the various principles of thermochemistry by explaining the concepts of calorimetry and its significance to thermochemistry. The chapter also discusses the various laws governing and related to thermochemistry.

The third chapter tackles the concepts related to structure and reactivity of ions with regard to thermochemistry in relation to the ion formation and the calculation of the energies related to them. The fourth chapter takes into account the kinetic molecular theory of gases and its applications in the thermochemical perspective.

The fifth chapter discusses in general the ways in which thermochemistry is affiliated with various chemical substances, while chapter six explains the thermochemistry of materials processing.

The principles of thermochemistry was also compiled in the various software and digital tools so that the analysis of the action of various forces and heat energies can be done on online platforms. The seventh chapter educates the readers on computational thermochemistry and how it can be used to analyse based on the various forces and external heat exchanges that impacts the system. The book then concludes by explaining the various applications of thermodynamics and providing rationale of the subjects significance to the field of chemistry and thermodynamic science.

With this book it is the author's goal to provide readers essential and relevant information that will be instrumental in understanding the various concepts, processes and applications related to thermochemistry in a systematic manner.

# Introduction to Thermochemistry

## CONTENTS

Thermochemistry is concerned with the exchange of heat energy that takes place during chemical reactions and the changes in state of a substance. This chapter introduces of thermochemistry in general and the various concepts associated with it.

The chapter first explains the scope of thermochemistry and the process of combustion then followed by brief recall of the history of thermochemistry. Some important concepts in thermochemistry included in this chapter are the internal energy, calorimetry, Hess' law, and standard enthalpies formation.

The chapter also explains thermochemical reactions, or reactions that are accompanied by absorption or release of heat. The chapter then explains other important aspects of thermochemistry such as thermodynamic systems, changes in internal energy, standard states, entropy, and Gibbs function. Some real-life examples of thermochemistry principles are also explained.

## 1.1. INTRODUCTION

The heart of industrial processes is knowing it from the most basic and simple to the most complex chemistry. The complexity of the process is dictated by the different basic sciences that governs it. In the industrial setting, both theory and practice are being collaborated to achieve desirable outcomes. Undoubtedly, that much has been wrapped up by more or less empirical methods. As a matter of fact, "theory" lags a lot behind "practice" in certain cases.

In spite of everything, this is not an argument for reducing the importance of theoretical principles to the background. When an industry is in an undeveloped state, empiricism is unavoidable and is ultimately the greatest barrier to further progress and development. In the widest sense, modern synthetic chemistry involves much more than being completely descriptive.

The success of a chemical process depends if there is enough energy for the process to take place and if the process proceeds at the expected rate. The exposition of all of these conditions should not, be considered merely a matter of trial and error, since chemical processes does not proceed this way. Considerations of a wide as well as general nature applicable to the processes of the most varied kind determine the rational control of such process.

The challenges of yield and efficiency, as well as the rate of reaction, are intimately bound up with such general considerations like its dependence to temperature, concentration, and pressure, together with the consideration of the equilibrium state as depicted by the equilibrium constant

In short, the technical applications of the principles of physical chemistry are represented by several challenges. Hence, we can conclude that it is important for chemists and/or chemical engineers to be adequately familiar if not adept with the principles of chemical thermodynamics and chemical kinetics.

## 1.2. THERMOCHEMISTRY

Thermochemistry is defined as the branch of chemistry concerned with the quantities of heat evolved or absorbed during chemical reactions.

The wide array of definitions of thermochemistry sums up to that "thermochemistry is the study of energy changes accompanying chemical as well as physical reactions".

To expound it a bit and differentiate with thermodynamics or chemistry, it is the study of heat energy which is associated with the physical transformations and/or chemical reactions governing a specific process. By nature, energy may be released or absorbed by a reaction or a system. Thus, as energy is release or absorbed by the system, this makes significant changes to the chemical and physical make up of a system.

These physical or chemical changes are assessed visually or chemically measured or both. A classic example is the phase change that occurs when a substance melts or boil. In terms of melting of ice, these phenomena will not take place if there is no enough chemical energy to allow the separation of solid atoms of water molecules, forming liquid flowing water. Thermochemistry pays attention on all these energy changes, specifically on the energy exchange of system with regard to its surroundings.

Thermochemistry has been proved beneficial in predicting product and reactant quantities from beginning to end of the course of a given reaction. It is also used to predict whether a reaction is favourable or unfavourable, spontaneous or non-spontaneous in combination with entropy determinations.

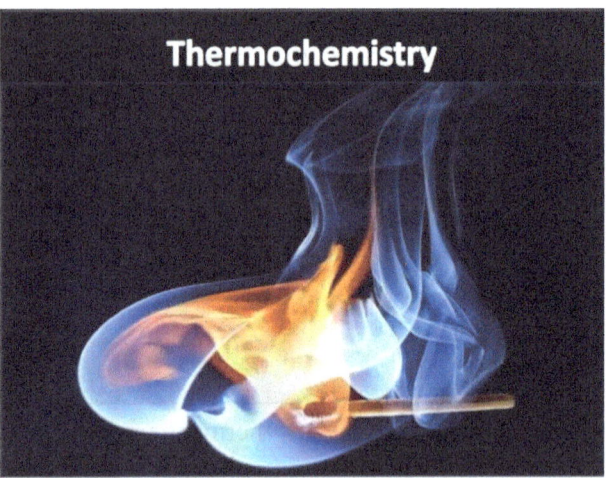

**Figure 1.1**. Thermochemistry is mainly concerned with absorption and release of energy.

*Source: Image by Wikipedia Commons.*

Thermodynamics is the study of the relationship between heat, work, and other forms of energy. The ability of a system to absorb and release heat is governed by the thermodynamics of the system. Thermochemistry is a branch of thermodynamics which focuses on the study of heat given off or absorbed in a chemical reaction. As such, it is a combination of the concepts of thermodynamics together with the concept of energy governing chemical bonds.

The subject generally involves calculations of such quantities as heat of combustion, heat capacity, heat of formation, entropy, enthalpy, free energy as well as calories.

Thermochemistry also involves the study of the heat capacities, entropies, and enthalpies of substances as well as paying attention to the physicochemical attributes of the systems such as the temperature dependence of these quantities and the corresponding phase changes taking place in the process . It is the specific area of chemical thermodynamics concerned with the calculation as well as measurement of the heat effects of reactions, and other heats associated with the reaction such as latent heat of vaporization and heat of formation.

Bonds among atoms are broken and then reformed when a reaction takes place. To break bonds, energy is required and when they are formed, energy is released. Generally, this takes place in the form of heat. Different reactions possess varied proportion of used energy to released energy. It concludes whether it is exothermic (releases more energy than it takes in) or endothermic (takes in more energy from its surroundings than it releases).

To quantify the heats released or absorbed by the system, calorimeters are used. A calorimeter is a device usually made up of a metal with good conductivity and a thermometer (or thermal sensor) that is sealed to prevent absorption or dissipation energy while measuring the heat capacity of the material. The science of energy and heat measurement using this device is known as calorimetry, which is a very important division of thermochemistry.

In order to predict the macroscopic behaviour of the system under study, chemical thermochemistry or thermodynamics developed into a powerful tool for chemical engineers, metallurgists as well as materials scientists. Thus, thermochemistry also involves the study of the amount of energy required to undergo any kind of physical or chemical transformation.

Process engineers generally look for phase changes (fusion or solidification) or for specific chemical reactions (oxidation, reduction, etc.), in materials processing. However, in all the cases, it is necessary to figure out the energy exchange involved in these changes. In order to keep up in any of these transformations, the importance of quantifying the energy is required for establishing energy and mass balances. This helps in sizing up the different equipment required to produce the materials that we end up using every day.

In the years 1752–1754, M. V. Lomonosov pointed out the importance of studying heat effects as well as heat capacities. He was recognized as the first one to explore this chemical process parameter. In the second half of the 18th century, J. Black, A. Lavoisier together with P. Laplace carried out the first thermochemical measurements. In the 19th century, calorimetric measuring techniques were refined by the hard work of scientists named as H. P. Thomsen, G. I. Gess (G. H. Hess), P. Berthelot and V. F. Luginin and significant progress in thermochemistry was marked in the early 20th century.

On the other hand, by the increase in temperature and accuracy range of experiments and as well as by the establishment of links among the energy effects of processes, the structure of particles (molecules, atoms, ions), including the position of elements in periodic system was conceived by D. I. Mendeleev. These tales place simultaneously with the remarkable increase in the number of substances that were studied and investigated. The theories on thermochemistry then became statistically-based combined with concepts from quantum chemistry in the mid-20th century.

Heat is not always measured directly. The difficulty as well as the impossibility of directly measuring the effects of heat of many processes often necessitates determinations using indirect methods. Reactions involving chemical processes sometimes become too complex such as multi-step reactions that makes measurement of heat and energy on each step not feasible. Thus, indirect measure of energy as well heat by electron affinities, lattice energy of ionic compounds, change in phase transitions and heat of formation of unstable intermediates are considered, as stated by the Hess Law. The Hess law is considered therefore as one of the most important underlying principle of thermodynamics and thermochemistry, stating the energy change in an overall chemical reaction is equal to the sum of the energy changes in the individual reactions comprising it.

For the reactions of organic compounds, standard heats of combustion are used. Using Kirchhoff's equation, heats of formation of chemical reactions may be recomputed at variable temperatures. It is often essential to resort to estimate relationships when the data needed for calculation are lacking. On the basis of composition and structure, as well as by analogy with previously studied substances and processes, these relationships make the measurement and determination possible of the several energy properties of processes and substances.

Data from experimentally established relationships as well as thermochemical studies are used to investigate the heat value of fuels, calculate chemical equilibrium, calculate the heat balance of technological processes, and determine the correlations between energy properties of substances and their composition, stability, structure, and reactivity of the substances. Thermochemical data together with other thermodynamic properties helps in the selection of the optimal conditions for the production of chemicals.

In the thermochemistry of solutions, broad advances have been made in determining heats of solution, mixing, heat capacities as well as vaporization, and the dependence of these quantities on concentration and temperature. The properties of individual components may be found, and the heats of solvation and heat effects of other processes may be calculated when these quantities are known. This information serves as basis for theories on the structure and nature of solutions. Thermochemical methods are also employed in some other fields, for instance, on the study of biological processes and detailed colloid chemistry, while one of the major uses of thermochemistry is for chemical engineers in design manufacturing plants, whereas biochemists use thermochemistry to understand bioenergetics.

The conversion of a set of substances collectively referred to as "reactants" to a set of substances collectively referred to as "products," are involved in chemical reactions. The reactants are gaseous methane, $CH_4$ (g), and gaseous molecular oxygen, $O_2$ (g), and the products are gaseous carbon dioxide, $CO_2$ (g), and liquid water $H_2O(l)$ in the balanced chemical reaction mentioned as follows:

$$CH_4(g) + 2\ O_2(g) \rightarrow CO_2(g) + 2\ H_2O(l) \qquad (1)$$

## 1.2.1. COMBUSTION

Combustion reactions are the reactions in which a fuel combines with oxygen to produce $H_2O$ and $CO_2$. A classic example of combustion reaction is the conversion of gaseous methane, $CH_4$ (g), and gaseous molecular oxygen, $O_2$ (g), forming gaseous carbon dioxide, $CO_2$ (g), and liquid water $H_2O(l)$. The balanced chemical reaction is presented as:

$$CH_4(g) + 2\ O_2(g) \rightarrow CO_2(g) + 2\ H_2O(l) \quad (1.1)$$

It is expected that reaction (1) will liberate heat because natural gas consists primarily of methane. Exothermic reactions are known as the reactions that liberate heat, and endothermic reactions are known as the reactions that absorb heat.

The heat related to a chemical reaction relies on the temperature and pressure of the reaction. All reactions, is assumed to be taking place at standard conditions (STP), or at temperature of 298 K (24.85°C) and applied pressure of one (1) bar. The amount of material undergoing reaction provides the basis of the quantity of heat released in a reaction.

In a reaction, the amount of reacting materials (reactants) are presented in the mole unit. The coefficient of the reactants is also equal to its amount in moles. In reaction (1) it these is 1 mole of methane having a mass of 16 grams (0.56 ounces), and requires 2 moles of $O_2$ (g) to produce the corresponding product of 1 mole gaseous of $CO_2$ and 2 moles of liquid water. The physical state of the products and reactants also provides the basis to thermochemistry. At STP, say the heat liberated by reaction 1 is 890 kilojoules (kJ). The reaction is written as:

$$CH_4(g) + 2\ O_2(g) \rightarrow CO_2(g) + 2\ H_2O(l) \qquad -\ 890\ kJ \qquad (1.1.1)$$

However, in case, $H_2O$ in the gas phase is formed, $H_2O(g)$, the heat released is just 802 kJ.

$$CH_4(g) + 2\ O_2(g) \rightarrow CO_2(g) + 2\ H_2O(l) \qquad\qquad - 802\ kJ \qquad (1.1.2)$$

Now, reversing reaction (1) , yields a reaction wherein 890 kJ of heat must be supplied for the reaction to take place. This mechanism is now written as follows:

$$CO_2(g) + 2\ H_2O(l) \rightarrow CH_4(g) + 2\ O_2(g) \qquad +890\ kJ$$

Several examples of exothermic reaction are: any form of neutralization, combustion (imagine of the heat released when you burn fuel), and most oxidation reactions (Figure 1.2).

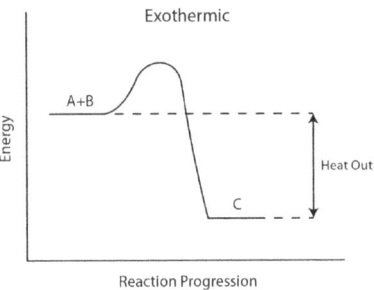

**Figure 1.2.** Exothermic reactions release energy.

*Source: Image by Wikimedia Commons.*

While examples of endothermic reactions are: decomposition, evaporation, and electrolysis. Therefore, the study of these processes as well as the factors involved is termed as thermochemistry.

## 1.3. HISTORICAL FACTS OF THERMOCHEMISTRY

The history of thermochemistry, as expected is well related and adhering to the history of thermodynamics. In the early 17th century, the term "caloric theory" was conceived stating that ice would melt only when if it was held in contact with a hotter body. The hotter body is expected to release a caloric particle into the ice that will enable it to melt. Later, Robert Hooke state that "heat is property of a body arising from the motion or agitation of its parts".

In 1761, Joseph Black, a Scottish physicist and chemist performed precise measurements of heat, stating that "addition of heat into ice at its melting point or to water at its boiling point does not result in change of temperature"

Later in the year 1789, Antoine Lavoisier as well as Pierre-Simon Laplace, a scientist and a mathematician, respectively, laid down the principles of thermochemistry. They both agreed and stated that the heat involved in a reaction is equal to the heat absorbed in the reverse reaction. Moreover, they also investigated latent heats, specific heats and amounts of heat involved in combustion of several substances.

In their experiment, they used a calorimeter to determine the heat evolved per unit of CO2 produced. This experiment is designed to measure the heat evolved in the respiration process of Guinea pigs. As a result, they found that animal respiration is a type of combustion reactions and that energy is produced during respiration.

The first truly quantitative chemical experiments include the experiments of Lavoisier. He minutely observed the reactants as well as products involved which is considered as a major step in the advancement of chemistry. Even though the matter proved by him changed its state, it has the same quantity in the beginning as well as the end of a chemical reaction.

The studies on the relationship between heat, energy and temperature progresses and in 1841, the turning point of thermodynamics and thermochemistry was when James Prescott Joule introduced the numerical value of the mechanical equivalent of heat which is 4.184 Joules per calorie which is also defined as the amount of work required to raise the temperature of 1 pound of water by 1 Fahrenheit.

Then Swiss chemist Germain Hess in the mid-18th formulated the very famous Hess Law, which is the framework of all energy and heat profiles of any chemical reaction. According to him, the evolution of heat in a reaction is the same whether the process is accomplished in multiple steps or one step. The total enthalpy during the course of a chemical reaction is the same no matter how many intermediate steps are between the initial and final step. The enthalpy or the until now is the basis of all evaluations of the standard heat of formations of direct synthesis calorimetry.

## 1.4. IMPORTANT CONCEPTS IN THERMOCHEMIS-TRY

*Work and Energy.*

There were two basic types of energy that plays a vital role in understanding and applying the principles of thermochemistry. Potential energy (PE) or stored energy is the energy of the substance by virtue of its position. Mathematically written as PE = mgh, PE is influenced by the height or displacement of an object as related to its mass and the gravitational force of energy. The other type of energy is the Kinetic Energy (KE) or the energy of an object by virtue of its motion. The formula KE = 1/2 mv2 presents the influence of the objects mass and velocity on kinetic energy.

On the other hand, work is done when there is change in kinetic or potential energy. As such, based on the Newton's 2nd Law of motion, we can relate work with energy by equations: W=ΔKE or W = -ΔPE. Similarly, work is defining as the product of force and displacement, W = Fs, where F is the force applied and s is the displacement. The SI unit for work and energy is Joules (N·m of kg·m2/s).

## The system and the surroundings

Energy s transferred from the surrounding to the system or vice versa. Substances act as reservoirs of energy *(system)*, meaning that energy can be added to them or removed from them. Parallel to this, the media the absorbs the energy when the system releases it, or provides the energy when the system needs it is called the *surroundings*. The transit if heat and energy to the system to the surroundings or vice versa. When energy is stored in a substance, the kinetic energy of its atoms or molecules is raised. When the system undergoes change, say an increase in kinetic energy, the energy is then transferred to the surroundings. Similarly, energy can be transferred from the surroundings to the system. When energy is transferred to the system (*q*) the system absorbs heat. When the energy is transferred into the surrounding, work (*w*) is done by the system. The changes in energy of the system then affects the system's internal energy (U or E).

Further, systems are classified to be an open system or a system that allows transit of heat of energy. A closed system, on the other hand allows only the transit of energy but not heat, while an isolated system does not allow the transit of both heat and energy.

## 1.4.1. Internal Energy (U), State Functions, and the First Law of Thermodynamics

Internal energy is denoted by letter U, while some textbooks denote it as E. It is the total of all the possible kinds of energy present in a substance. It is a state function. In simple terms, it has a unique value once the state or condition of a system is defined. The difference between the internal energy of the final and initial states is equal to the change in the internal energy of a system ($\Delta U$ = $\Delta U2 - \Delta U1$), which also obeys the law of conservation of energy.

Though energy cannot be created nor destroyed, it can be transferred from or to the surroundings. The first law of thermodynamics ($\Delta U = q + w$) explains the relationship between change in the internal energy of a system and the heat and work exchanged between a system and its surroundings.

## 1.4.2. Heats of Reaction and Enthalpy Change $\Delta H$

The heat of reaction is the quantity of heat exchanged between the reaction system as well as its surroundings, for any reaction occurring at constant temperature. The heat of reaction at constant volume ($qv$) is equal to the change in internal energy ($\Delta U$), if a reaction is carried out at constant volume. Enthalpy ($H$) is a more useful property than internal energy for constant-pressure processes to which it is related. The heat of reaction ($qp$) is equal to the enthalpy change ($\Delta H$) at a constant pressure:

$$qp = \Delta H = \Delta U + P\Delta V \qquad (1.2)$$

Like internal energy, enthalpy is a state function.  . Enthalpy changes can be written into chemical equations. It can also be incorporated into conversion factors that relate the amounts of substances to quantities of heat absorbed or released in chemical reactions. Enthalpy decreases and heat is given off to the surroundings in an exothermic reaction on the other hand enthalpy increases and heat is absorbed from the surroundings in an endothermic reaction.

## 1.4.3. Calorimetry

The process of measuring quantities of heat is called calorimetry and a calorimeter is an instrument used for the procedure. Masses and temperature changes are the actual measurements. Heat capacities and/or specific heats are the other data required. The quantity of heat needed to change the temperature of the system by 1°C (or by 1 K) is the heat capacity (C) of a system:

$$C = \frac{q}{\Delta T} \qquad (1.3)$$

The heat capacity of one mole of substance is known as the molar heat capacity. The heat capacity of a 1-g sample of the substance is called specific heat:

$$\text{Specific heat} = \frac{\text{heat capacity}}{\text{mass}} = \frac{C}{m} = \frac{q}{m \times \Delta T} \qquad (1.3.1)$$

At constant pressure of the atmosphere in a plastic-foam calorimeter, a number of reactions can be carried out. The *qp* or $\Delta H$ values are the results obtained. The reactions which are generally carried out under constant-volume conditions in a bomb calorimeter are still combustion reactions. However, work does not expand in a constant-volume environment. Thus, qv or $\Delta U$ values must be obtained in terms of qp. It is important to note that $\Delta H$ values can be calculated from $\Delta U$ values when necessary.

## 1.4.4. Hess's Law of Constant Heat Summation

The total enthalpy during the course of a chemical reaction is the same no matter how many intermediate steps are between the initial and final step. Similarly, the change in enthalpy $\Delta H$ is the same whether a reaction is carried out in one step or several because enthalpy is a state function. We can use Hess's law to evaluate $\Delta H$ by summing the enthalpy changes of the steps and write the equation for a reaction as several steps.

To simply understand Hess law, Figure 1.3 depicts 3 step wise process as:

$$\Delta H = \Delta H1 + \Delta H2 + \Delta H3 \qquad (1.4)$$

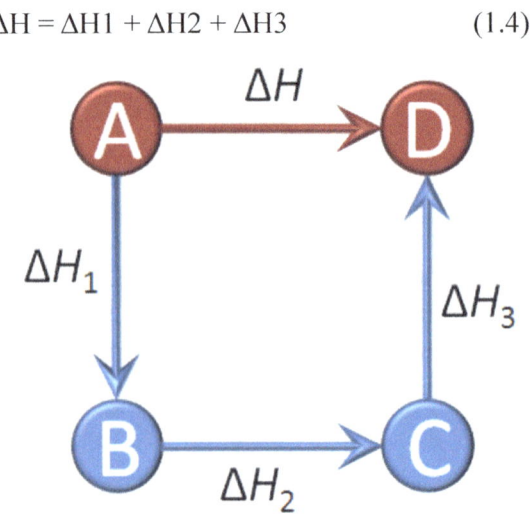

**Figure 1.3**. Representation of Hess's Law.

***Source:*** *Image by Wikimedia Commons.*

## 1.4.5. Standard Enthalpies of Formation

The standard state of a pure liquid or solid is at 1 atm whereas for a gas, it is the pure gas behaving ideally at 1 atm at the temperature of interest. The enthalpy changes of a reaction in which both the reactants as well as the products are in their standard states is called their standard enthalpy of reaction, $\Delta H°$. The standard enthalpy of formation, $\Delta H°f$ is equivalent to zero for an element in its reference form. The standard enthalpy changes of a reaction in which the compound is formed from its elemental as their standard state is the standard enthalpy of formation of a compound. To calculate standard enthalpies of reaction through the following equation (where the symbols ($vr$) and ($vp$) are stoichiometric coefficients), standard enthalpies of formation (usually listed at 25.00°C – standard temperature) can be used:

$$\Delta H° = \sum v_p \times \Delta H°_f (\text{products}) - \Delta H°_f (\text{reac tan ts})$$

(1.5)

## 1.4.6. Combustion and Respiration: Fuels and Food

Two aspects of life in which thermochemical principles play key roles are explored in in terms of; metabolism of foods—carbohydrates, fats as well as proteins, and the combustion of fossil fuels—petroleum, natural gas as well as coal.It is an established fact that heat is either released or absorbed in chemical reactions. Endothermic reactions are those reactions that absorb heat and exothermic reactions are those reactions that release heat. The examples of endothermic reactions are vaporization of H2O, sublimation of naphthalene, solvation of sugar in H2O. In an endothermic reaction, the potential energy of reactants is lower than potential energy of products in endothermic reactions. Heat is then given to reaction to balance this energy difference (see Figure 1.4).

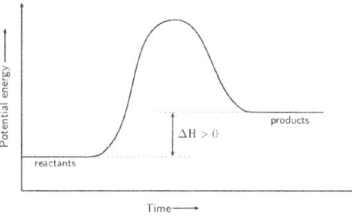

**Figure 1.4.** Energy diagram of an endothermic reaction

***Source:*** *https://socratic.org/questions/how-can-i-represent-an-endothermic-reaction-in-a-potential-energy-diagram*

In endothermic reactions H2 is always larger than H1, as you can see in the given graph. Hence, ΔH=H2-H1 is always positive or favourable. We write it like the following in the reactions:

$CaCO_3(s)$ + Heat → $CaO(s)$ + $CO_2(g)$  (+178 kJ)          (1.6)

This equation is written correctly as:

$CaCO_3(s)$ → $CaO(s)$ + $CO_2(g)$          (+178 kJ)(1.6.1)

When heat becomes part of the reactants, this means that heat must be supplied to the system, indicating an endothermic reaction.  The resulting enthalpy of formation of $CaCO_3$ was then calculated to be positive 178 kJ.

For exothermic reactions, condensation of gases and combustion reactions are two popular examples of how heat is liberated by a chemical reaction. As discussed earlier, potential energies of the reactants are higher than potential energies of products in these reactions. To balance energy difference, the excess amount of energy is written in right side of reaction.

$H_2O(g)$ → $H_2O(l)$ + Heat                    (1.7)

$O$ + e- → $O^{-1}$ + Heat                    (1.7.1)

$H_2(g)$ + ½ $O_2(g)$ → $H_2O(g)$ + Heat          (1.7.2)

This means also that when heat is found to be at the product side of the reaction, heat is given off and the reaction is exothermic.

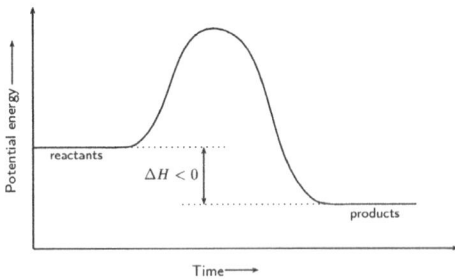

**Figure 1.5.** Energy Diagram of an exothermic reaction

*Source:     https://www.siyavula.com/read/science/grade-11/energy-and-chemi-cal-change/12-energy-and-chemical-change-02*

In exothermic reactions H1 is always larger than H2, as you indicated in Figure 1.5. Therefore, $\Delta H = H2 - H1$ is negative or unfavourable. We write it like the following in reactions;

$N2(g) + 3H2(g) \rightarrow 2NH3(g) + Heat$              (-46 kJ)              (1.8)

This equation therefore is written correctly as:

$N2(g) + 3H2(g) \rightarrow 2NH3(g)$              -46 kJ
(1.8.1)

Example: Which one of the following reactions is exothermic ($\Delta H$ is negative)?

I. $H2O(g) \rightarrow H2O(l)$    $\Delta H1$

II. $K(g) \rightarrow K+(g) + e-$    $\Delta H2$

III. $C(s) + O2(g) \rightarrow CO2(g)$    $\Delta H3$

Solution

The energy is released when matters change state from gas to liquid. It is an exothermic reaction and $\Delta H1$ is negative.

We should give energy to remove one electron from atom. So, II is endothermic reaction. Moreover, $\Delta H2$ is positive.

Energy (heat) is released in combustion reactions. So, III is exothermic reaction and $\Delta H3$ is negative.

# 1.5. THERMOCHEMICAL EQUATION

The reaction between methane gas and oxygen is exothermic, because it is primarily a combustion reaction. Specifically, 890.4 kJ of heat energy are released by the combustion of 1 mol of methane. This is shown as part of the balanced equation:

$CH4(g) + 2O2(g) \rightarrow CO2(g) + 2H2O(l) + 890.4$ kJ              (1.9)

The equation states that when 1 mol of methane is combined with 2 mol of oxygen, we get 1 mol of CO2 and 2 mols of H2O and 890.4 kJ of heat is released in the process and thus, it is written as a product of the reaction. A chemical equation that involves the enthalpy change of the reaction is known as a thermochemical equation. Given below is the figure which explains the process in the above thermochemical equation:

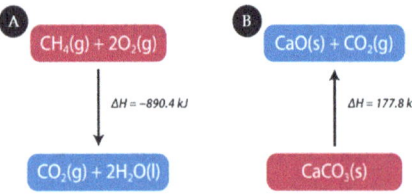

In an exothermic reaction, as reactants are converted to products, enthalpy is released into the surroundings. In A, since heat is released, the enthalpy change of the reaction is found to be negative. In B, the enthalpy is absorbed from the surroundings as reactants are converted to products. This reaction is then classified as an endothermic reaction and the enthalpy change of the reaction is found to be positive.

In the combustion of methane, heat is being released by the system so the enthalpy change is negative. Hence, there is decrease in overall enthalpy of the system. The enthalpy change for a chemical reaction is the heat of reaction. The heat of reaction is now found to be −890.4 kJ for the combustion of methane. Thermochemical reaction can also be written in the following way:

$$CH4(g)+2O2(g)\rightarrow CO2(g)+2H2O(l) \qquad \Delta H=-890.4 \text{ kJ} \qquad (1.9.1)$$

The heat of reaction is usually expresses in kilo Joules or simply kJ. It is necessary to involve the physical states of the reactants as well as products in a thermochemical equation because the value of the $\Delta H$ relies on those states.

The decomposition of calcium carbonate to calcium oxide and carbon dioxide requires heat.  Thus, the reaction is considered an endothermic reaction.  177.8 kJ of heat must be absorbed for 1 mol of calcium carbonate to decompose into 1 mol of calcium oxide and 1 mol of CO2.. The thermochemical reaction is shown as follows:

$$CaCO3(s)+177.8 \text{ kJ} \rightarrow CaO(s)+CO2(g \qquad\qquad (1.10)$$

The 177.8 kJ is written as a reactant because the heat is absorbed by the system. For an endothermic reaction, the heat of reaction is positive.

$$CaCO3(s) \rightarrow CaO(s)+CO2(g) \qquad \Delta H=177.8 \text{ kJ} \qquad (1.10.1)$$

The value of the enthalpy change for the reaction is influenced by the way in which the  reaction is written. A number of reactions are reversible in nature. It means that the product(s) of the reaction are capable of combining as well, reforming the reactant(s). The sign of the $\Delta H$ changes when a reaction is written in the reverse direction. For instance, if you allow calcium oxide to react with carbon dioxide, calcium carbonate forms together with the liberation of heat. This is the reverse reaction of the decomposition of calcium carbonate discussed earlier, Thus, the thermochemical reaction above is also reversed. And the reaction is now considered as exothermic:

$$CaO(s)+CO_2(g) \rightarrow CaCO_3(s)+177.8 \text{ kJ} \qquad (1.10.2)$$

$$CaO(s)+CO_2(g) \rightarrow CaCO_3(s) \qquad \Delta H=-177.8 \text{ kJ}....(1.10.3)$$

## 1.5.1. Stoichiometry of Thermochemical Equations

Chemical equation for a reaction with the enthalpy of reaction shown immediately after the equation is a thermochemical equation.

$$H_2(g) +1/2 O_2(g) \Rightarrow H_2O(l) \ \Delta H =-285.840 \text{ kJ} \qquad (1.11)$$

You will find the following rules helpful while working with thermochemical equations:

The thermochemical equation should include phase labels because the change in enthalpy will change if the phases are different:

$$H_2(g) + \tfrac{1}{2} O_2(g) \Rightarrow H_2O(l) \qquad \Delta H = -285.840 \text{ kJ} \qquad (1.11.1)$$

$$H_2(g) + \tfrac{1}{2} O_2(g) \Rightarrow H_2O(g) \qquad \Delta H = -241.826 \text{ kJ} \qquad (1.11.2)$$

The magnitude of $\Delta H$ is proportional to the amount reacting substances

$$H_2(g) + \tfrac{1}{2} O_2(g) \Rightarrow H_2O(l) \qquad \Delta H = -285.840 \text{ kJ} \qquad (1.11.3)$$

285.840 KJ is thermodynamically equivalent to 1 mol of $H_2$

285.840 KJ is thermodynamically equivalent to 1/2 mol of $O_2$

285.840 KJ is thermodynamically equivalent to 1 mol of $H_2O$

Now, say you want to remove coefficients that are in decimals, the equation above must be multiplied by 2:

$$[ H_2(g) + \tfrac{1}{2} O_2(g) \Rightarrow H_2O(l)] \ x \ 2 \quad \text{which gives:} \qquad (1.12)$$

2 $H_2(g) + O_2(g) \Rightarrow 2 H_2O(l)$ , then its original $\Delta H$ must also be multiplied by 2, thus:

2 $H_2(g) + O_2(g) \Rightarrow 2 H_2O(l) \quad \Delta H = 2(-285.840)$ kJ or -571.68 kJ (1.12.1)

Therefore:

When a thermochemical equation is multiplied by a factor, the value of $\Delta H$ for the new equation is obtained by multiplying by the value of $\Delta H$ by the same factor:

$H_2(g) + \frac{1}{2} O_2(g) \Rightarrow H_2O(l)$ $\qquad \Delta H = -285.840$ kJ $\qquad$ (1.13)

$2H_2(g) + O_2(g) \Rightarrow 2H_2O(l)$ $\qquad \Delta H = -= 571.68$ kJ $\qquad$ (1.13.1)

And, when a chemical equation is reversed, the sign of $\Delta H$ is also reversed:

$H_2(g) + \frac{1}{2} O_2(g) \Rightarrow H_2O(l)$ $\qquad \Delta H = -285.840$ kJ $\qquad$ (1.14)

$H_2O(l) \Rightarrow H_2(g) + \frac{1}{2} O_2(g)$ $\qquad \Delta H = +285.840$ kJ $\qquad$ (1.14.1)

## 1.5.2. Factors that Control Reactions

The factors that affect the speed of the chemical reactions is categorised into: kinetic and thermodynamic factors. Kinetics deal with the speed at which a reaction takes place while thermodynamic concepts deal with the energetics of a system.

The mechanism of the reaction is related with observations of reaction kinetic. In short, kinetics will tell you how fast the reaction proceeds and thermodynamics tell you is the reaction will proceed or not.

This explains the manner in which we assume that the atoms as well as molecules react during a reaction. In order to indicate that the reactants take part in a reaction that leads to the formation of the products and that the process goes to completion, we often write equations for chemical reactions with a forward arrow (e.g., Eq. 2.1).

$Zn(s) + H_2SO_4(aq)$ $\qquad ZnSO_4(aq) + H_2(g)$ $\qquad$ (1.15)

However, it is a fact that many reactions do not reach at the stage of completion. Alternatively, reactants as well as products remains in equilibrium in which both forward occurs together with the reverse or backward reaction. Surprisingly, all reactions are in equilibria. In other words, no reaction under equilibrium conditions goes completely to the right-hand side.

Thermodynamic factors govern the position of equilibrium. The sign and magnitude of the change in Gibbs energy, $\Delta G$, are used to assess whether a reaction is favourable and to what extent it will reach completion. Strictly, information about the favourability of a reaction is given by the change in Gibbs energy. Thermochemical data can also be used to gain some insight, i.e. the changes in heat that accompany chemical reactions. By measuring the associated change in temperature, the heat change that accompanies a reaction can be readily determined.

. In the current chapter, we study the changes in heat (enthalpy), not only for chemical reactions, but for phase transitions as well. The enthalpy terms that are associated with interactions between molecules are also been introduced in the previous sections.

## 1.6. GENERAL OVERVIEW OF THERMOCHEMISTRY

Energy is defined as the capacity to perform work or transfer heat. Kinetic Energy (KE) is defined as the energy of motion:

$$E_{Kinetic} = KE = \frac{1}{2}mv^2$$

(1.16)

In the KE equation above, m corresponds to the mass of the object while v is the velocity of the object. Simply, we take KE as the energy of moving objects. It is important that the object is the one the moves from one point to another.

The energy that a system possesses by virtue of its position or composition is then called potential energy. In the formula below, m corresponds to the mass of the object, g is the gravitational force of the location and h is the elevation. In the concept of PE, the gravitational force is classified to be locational. This means that its value changes depending on the object's location, like Earth and Mars. The gravitational force of the earth differs to that of Mars, which means that the potential energy of an object in Mars with the same mass and elevation will be 6x of its value when the object is in Earth:

$$E_{Potential} = PE = mgh$$

(1.17)

The form of energy that always flows spontaneously from a hotter body to a colder body is heat. The measurement of heat in terms of the temperature from its initial to its final point. Temperature is also the property of a material that established the relationship between heat and energy.

### 1.6.1. Thermodynamic Systems

The system is the substance that is being considered in a study and it has a boundary. Thus, everything outside the boundary is called its surroundings. Thus, the transit of matter or energy or both into the system to the surroundings and vice versa is being limited by the boundary. A set of conditions that completely specify all the properties of the system is known as the thermodynamic state of that system. This set of condition involves the temperature (T), volume (V) and composition, pressure (P) as well as physical state (gas (g), liquid (l) or solid (s)).. Sate functions are known as the properties of a system such as P, V, and T. All the energy contained within the system is represented by the internal energy of that system. It involves the energies of attraction, kinetic energies of the molecules and repulsion forces among subatomic particles, ions, atoms, and molecules(and other forms of energy).

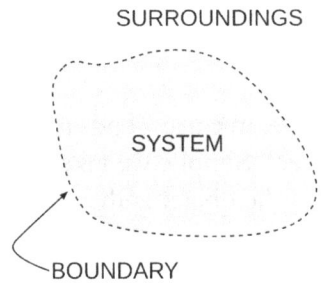

**Figure 1.4.** Thermodynamic systems and their surroundings separated by boundary.

*Source: Image by Wikipedia.*

### 1.6.2. Changes in Internal Energy

From the First Law of Thermodynamics stating change in the internal energy of a system is equal to the sum of the heat gained or lost by the system (q) and the work done by or on the system (w):

$$\Delta E = q + w \qquad (1.18)$$

It is important to put in mind the universal sign convention of heat and work. Heat is negative if it is lost and positive if it is gained by the system. Similarly, work is positive when the work is done on the system and negative when is done by system.

In an isochoric or isometric process, the volume is constant, therefore w = 0. Hence, equation (1.18) becomes:

$$\Delta E = q(v) \qquad (1.18.1)$$

The amount of heat absorbed or released at constant volume is just the change in the internal energy.

At constant-pressure (isobaric) processes, the equation is as equation (1.18) is now driven with a change in volume term. The enthalpy is now based on a constant pressure system (qv) and is now written:

Changes into

$$\Delta E = qp - P\Delta V \tag{1.18.2}$$

As it undergoes a chemical or physical change at constant pressure, the quantity of heat transferred into or out of a system and qp, is defined as the enthalpy change, $\Delta H$, of the process.

$$qp = \Delta H = \Delta E + P\Delta V \text{ (constant T, P)} \tag{1.18.3}$$

## 1.6.3. Standard States

The most stable pure form under standard pressure (1 bar) and some specific temperature (298 K unless otherwise specified) is the thermodynamic standard state of a substance.

The standard state is the pure liquid or solid (C(graphite), H2O(l), CaCO3 (s)) for a pure substance in the liquid or solid phase. For a gas, the standard state is the gas at a pressure of one atmosphere. The standard state refers to one-molar concentration for a substance in solution.

The standard enthalpy change, $\Delta Hr$ for a reaction refers to the $\Delta H$ when specified moles of reactants are converted completely to the specified number of moles of products, all at standard states.

The enthalpy change from the reaction in which one mole of the substance in a specified state is formed from its elements in their standard states is called the standard molar enthalpy of formation (heat of formation), $\Delta H$. The $\Delta H$ value for any element in its standard state is zero by convention.

$$\frac{1}{2} H_2(g) + \frac{1}{2} Br_2(l) \rightarrow HBr(g) \tag{1.19}$$

$\Delta Hr = -36, 4$ kJ/mol      (this means that it takes -36.kJ/mol of energy to form HBr)

$\Delta Hf$ HB (g) = $-36, 4$ kJ/mol    (therefore, the heat of formation of HBR is -36 kJ/mol)

reactants $\rightarrow$ products

The sum of the standard molar enthalpies of formation of the products, minus the corresponding sum of the standard molar enthalpies of formation of the reactants is equivalent to the standard enthalpy change of the reaction.

$$\Delta H_r^\ominus = \sum \Delta H_f^\ominus \text{ products} - \sum \Delta H_f^\ominus \text{ reactants}$$

$$\tag{1.20}$$

Applications

1.    Say you have a reaction that absorbs 8 kJ of heat and does 2 kJ of work on its surroundings. What is the $\Delta H$ and $\Delta E$ od the system?

Solution:

From equation 1.18: $\Delta E = q - w = 0$; we denote w to be negative since it the system is the one that performed work and it was given off to the surrounding (work done by the system). Thus:
$\Delta E = q - w = 8 \text{ kJ} - 2 \text{ kJ} = 6 \text{ kJ}$

While $\Delta H = +qp$ ; we assume that heat is supplied at constant pressure.

$$\Delta H = +qp = +8kJ$$

2.    Using the reaction: 2 CO(g) + O2(g) • 2 CO2(g) $\Delta H = -566$ kJ

How much work is done and the change in internal energy when 9.0 g of CO reacts with oxygen at 1 atm and a change in volume of -3.2 L?

Solution:

Given:

2 CO(g) + O2(g) • 2 CO2(g)                         $\Delta H = -566$ kJ

m = 9.0 g

MW = 28 g/mol

Required: W, $\Delta E$

2.    Solution:

a)    For the pressure-volume work, from thermodynamics, at constant pressure:

W = -P$\Delta$V = -( 1 atm)(-3.2 L) = -3.2 L-atm

Using a cOnversion factor: 101.3 J = 1 L-atm;

$$W = 3.2l - atm \left( \frac{101.3J}{1 - atm} \right)$$
$= 324.16 \text{ J} = 0.324 \text{ kJ}$

b)    Using the formula for the change in internal energy:

$\Delta U = qp + W$

$$\Delta U = \left[ \left( \frac{9.0g}{2mols(28g \, / \, mol)} \right) (-566kJ) \right] + 0.324kJ$$

$\Delta U = -90.64$ kJ

## 1.6.4. Entropy

From thermodynamics, a spontaneous reaction is a reaction where:

$\Delta S(system) + \Delta S(surroundings) > 0$ (1.21)

while for a reversible process (a process proceeding only trough equilibrium states):

$\Delta Srev = qrev/T$ (1.21.1)

and when the process is irreversible:

$\Delta S > q /T$ (1.21.2)

Similarly, a constant volume irreversible process:

$\Delta U = qv \leq T\Delta S$ (1.21.3)

Predictable entropy ($\Delta S$) changes for the system are the results of processes: Phase changes (e.g., melting $\Delta S(system) > 0$, freezing $\Delta S(system) < 0$)., temperature, volume changes, mixing of substances, increase in the number of particles and changes in the number of moles of gaseous substances.

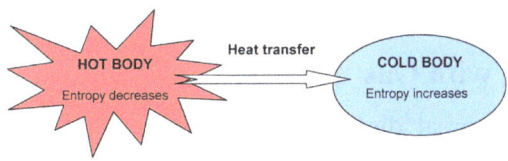

**Figure 1.5**. Heat transfer between two bodies causes a change in their entropy.

*Source: Image by Wikimedia Commons.*

### 1.6.5. Gibbs Function

At constant pressure:

$\Delta H = qp$

but:   $\Delta H \leq T\Delta S$

Gibbs free energy:

The indicator of spontaneity of a reaction or physical change at constant T and P is known as the Gibbs free energy. The process is spontaneous in case, $\Delta G$ is negative (the reaction favours product formation).

$\Delta G \leq 0$

$\Delta G < 0$ reaction is spontaneous (product formation is favoured)

$\Delta G > 0$ reaction is nonspontaneous (reactant formation is favoured)

$\Delta G = 0$ system is at equilibrium

$G = H - TS$

$\Delta G = \Delta H - T\Delta S$

Therefore, at constant pressure: The maximum useful energy obtainable in the form of work from a given process at constant temperature and pressure is the amount by which the Gibbs free energy decreases.

$\Delta H = q + w + P\Delta V$ (constant p)

At the time of a reversible process (constant T)

$q = T\Delta S$

$\Delta G = \Delta H - T\Delta S$

$\Delta G = T\Delta S + w + P\Delta V - T\Delta S$

## 1.7. EXAMPLES OF THERMOCHEMISTRY PRINCIPLES IN REAL-WORLD

### 1.7.1. Cooking with Gas

There has been a great discussion about how our country will meet its growing demand for energy for many years. Often, this discussion has focused on worries about the access to adequate supply of fuel and its consumption. More recently, the use of hydraulic fracturing (or "fracking") to release natural gas deposits trapped underground has been the topic of extensive debate.

The combustion of fossil fuels in for energy production remains an excellent context for teaching thermochemistry, although these discussions have often been politically charged. In order to produce heat and to perform work, energy is required. Natural gas is a fossil fuel Capable of storing energy as potential energy.

The major uses of natural gas includes heating food whether it is for preparation or consumption while compressed natural gas (CNG) are used by vehicles for public and mass transport. As compared to any of the other contexts presented, the operation of a vehicle allows the discussion of PV work in a more natural setting. Combustion of natural gas results in change of internal energy of the system, producing different amounts of heat and work, depending on conditions.

The burning of methane as seen on the stove produces mostly heat and does no work at all on the surroundings, whereas in the piston of a bus engine, releases heat and does work upon burning methane. However, the change in internal energy is a state function and on the other hand, heat and work are not. Heat released from exothermic combustion of natural gas at same pressure is equal to the enthalpy of combustion which can also convert $H_2O$ (to steam) for electric power generation. In this example, the transfer of heat illustrates the first law of thermodynamics.

Electric power generation using steam turbines is another completely different context. However, it permits discussion of the heating curve of $H_2O$, where the molar heat of vaporization and molar heat capacity are necessary and of great importance. One could also introduce heat capacity through a discussion in the context of cooking, like for example making pizza. Wolke states that brick and stone possess higher heat capacities as compared to metals so that they hold their temperature better, which now implies that heating value of a fuel is another factor important for fuels

According to the U.S. Energy Information Administration, in 2015, approximately 30% of the electricity produced in the United States was generated from the combustion of natural gas. This, using calorimetry or by applying Hess's law and standard enthalpies of formation, the enthalpy of combustion can be measured or computed.

## 1.7.2. Walden Pond

In order to introduce many of the principles of thermochemistry discussed in chapter 2, the seasonal changes of a pond (perhaps Henry David Thoreau's Walden Pond) can be served. However, it is difficult to discuss standard enthalpies of reaction in case the context is restricted to phase changes. However, one may start with the three phases of $H_2O$.

The changes at latitudes where the calendar year is defined by a cycle of four seasons, a freshwater pond provides a familiar context for introducing concepts of energy, and thermochemistry as well as thermodynamics. Ponds are covered with ice in winter. During spring, sunshine melts the ice, warms the $H_2O$ and ultimately leads to the spring bloom. As spring gives way to summer warmth, aquatic life reawakens.

In the cool early mornings of late summer, the fog and mist that form over $H_2O$ are the outcome of condensation of $H_2O$ vapor to a colloid of water droplets which foreshadows the return of winter cold together with the ice. The freezing of $H_2O$ during winter is an exothermic process, whereas the melting of ice during the spring is an endothermic process. Heat flows from the relatively warm surface (system) into the cold air (surroundings). As $H_2O$ vapor (the system) loses heat (q < 0) to its surroundings (i.e. the surrounding absorb heat from the system), the condensation of H2O vapor forming a mist or fog on the surface is an exothermic process whereas the evaporation of H2O from the surface of the pond is an example of endothermic process (q > 0).

The energy changes related with the phase changes of H2O require the use of the molar heat of fusion of ice, the molar heat capacity together with the molar heat of vaporization of water. Introducing the idea of work as a component of internal energy will require a digression from the example of a pond, as in several of the contexts for this chapter. In case the biological processes that sustain life in an aquatic environment is considered, it is possible to introduce the concepts of standard enthalpies of formation as well as the standard enthalpies of reaction into the discussion.

Most of the life forms in ponds relies on the combustion of glucose. On the other hand, glucose molecules are more difficult as compared to the simpler fuels. However, we can still measure the standard enthalpy of combustion of glucose using a calorimeter or compute such value using the Hess's law.

## 1.8. CONCLUSION

Thermochemistry studies the heat exchange that takes place during a chemical reaction. It is related to thermodynamics and is used widely by scientists and other experts like biochemists, chemical engineers and others in their respective fields.

Combustion reactions releases energy. In these reactions, fuel combines with oxygen to produce H2O and CO2. Methane is an important component in natural gas and once methane combusts, the reaction liberates heat. Reactions that releases heat are called exothermic reactions. Endothermic reactions are those that absorbs heat.

Thermodynamic systems are important concepts, which represents an area whose thermochemical changes are studied. The surroundings represent those areas which lie outside the system which is then separated by a boundary. The system is the reaction products and reactants and the surroundings comprise the entire area in the universe outside it.

Thermochemical equations represent a balanced stoichiometric chemical equation comprising of the enthalpy change ΔH value. Enthalpy refers to the transfer of energy in a reaction. In case of chemical reactions, heat exchange takes place.

$\Delta H$ represents the change in enthalpy and is a state function. This means that $\Delta H$ is independent of the initial and final states. In other words, $\Delta H$ will always be the same, irrespective of the steps taken from initial (reactants) to the final step (product/s formation).

The first law of thermodynamics is concerned with the relation between the enthalpy and the amount of heat and work that crosses the boundary of the system. Enthalpy is a state function. Hence, the heat released or absorbed during a chemical reaction is independent of whether the reaction proceeds from reactants to produce in a single or multi-step.

The Hess' law states that when two reactions combine together and yield a third reaction, the sum of the $\Delta H$os for the first two reactions is equal to the $\Delta H$o of the third.

# REFERENCES

1. Chemistryexplained.com. (n.d.). Thermochemistry – Chemistry Encyclopedia – reaction, Water, Gas, Symbol, Equation, Mass, Energy and Enthalpy. [online] Available at: http://www.chemistryexplained. com/Te-Va/Thermochemistry.html (accessed on 13 March 2020).

2. Chemistrytutorials.org. (n.d.). Thermochemistry | Online Chemistry Tutorials. [online] Available at: https://www.chemistrytutorials.org/ content/thermochemistry (accessed on 13 March 2020).

3. CK-12 Foundation, (n.d.). Thermochemical Equation. [online] Available at: https://www.ck12.org/book/ck-12-chemistry-concepts-intermediate/section/17.10/ (accessed on 13 March 2020).

4. LabRoots, (n.d.). Thermochemistry | Content Tag. [online] Available at: https://www.labroots.com/tag/thermochemistry (accessed on 13 March 2020).

5. Socratic.org. (2015). What is Thermochemistry? + Example. [online] Available at: https://socratic.org/questions/what-is-thermochemistry (accessed on 13 March 2020).

6. Stephanierushton.tripod.com. (n.d.). History. [online] Available at: http://stephanierushton.tripod.com/id1.html (accessed on 13 March 2020).

7. Szalai, I., (n.d.). Thermochemistry, thermodynamics. [ebook] Available at: http://nlcd.elte.hu/szalai/pdf/lecture-9-handout.pdf (accessed on 13 March 2020).

8. TheFreeDictionary.com. (n.d.). Thermochemistry. [online] Available at: https://encyclopedia2.thefreedictionary.com/Thermochemistry (accessed on 13 March 2020).

9. Thermochemistry, (n.d.). [ebook] Available at: http://catalogue. pearsoned.co.uk/assets/hip/gb/hip_gb_pearsonhighered/ samplechapter/Housecroft%20-%20chapter%202.pdf (accessed on 13 March 2020).

10. Thermochemistry: Energy Changes in Chemical Reactions, (n.d.). [ebook] Available at: https://wwnorton.com/college/nrl//temp/ chemat2_irm_ch09_20161025.pdf (accessed on 13 March 2020).

11. Wps.prenhall.com. (n.d.). Concept Review with Key Terms. [online] Available at: http://wps.prenhall.com/esm_hillpetrucci_ genchem_4/0,8603,1079637-,00.html (accessed on 13 March 2020).

# Principles of Thermochemistry

## CONTENTS

This section discusses the principles of thermochemistry: the various important concepts, laws, and principles related to thermochemistry. The chapter also provides discussion on heat transfer and temperature changes. It also provides techniques in determining specific heat capacity and its role in a chemical reaction. The chapter also throws some light on Hess's law. The meaning of calorimetry and how do calorimeters work is also tackled as well as the laws of thermochemistry, the meaning of state function and some other important also discussed.

## 2.1. INTRODUCTION

As we progressed over the basics of thermochemistry, everybody will agree that one needs a thorough understanding of the basics of thermodynamics and the necessary understanding of the relationship between heat, work, and energy and their transformations. When analysing chemical processes, it is very important to outline the system, the reaction/s or item of interest differentiate them from the surroundings, everything else.

An isolated system refers to a system in which neither matter nor energy can be transferred to or from the surroundings. A closed system refers to a system in which no matter or energy can be transferred to or from the surroundings. In mostly every cases in chemistry, there is a closed system of interest, and the energy and internal energy of the system, changes when energy in the form of heat (q) is lost or added and work (w) is done by the system or on the system, must be totally accounted for..

While the total amount of internal energy of a system cannot be directly measured, the alteration in the internal energy of the system, is calculated with the help of the equation discussed below:

$\Delta E_{system} = q + w$

- Where q refers to the energy that is in the form of heat exchanged between surroundings and system; and
- *w* refers to the work done by the system or on the system.

The first law of thermodynamics suggests that the total amount of energy for an isolated system is constant. This means that the total amount of energy and matter in an isolated system is constant. It is important to note that the energy can neither be created nor destroyed during any kind of physical changes, or chemical reaction, but it can transform from one form to another. Similarly, it can be said the energy is conserved during a physical or chemical change, or

$\Delta E_{universe} = 0$

(2.1)

As explained in the Figure 2.1. below, it clearly shows that the sign convention plays very important role in thermodynamics because of the fact that it shows what is happening to the internal energy of the system. When energy, in the form of heat is transferred from the surroundings to the system, q is positive, while on the other hand, heat the transfer is from the system to the surroundings, q is negative. In a similar way, when surroundings are responsible for doing the work on the system, w is positive, while it is negative when system is responsible for doing work on the surroundings.

Summarizing the sign conventions for heat and work in thermodynamics:

+q and +w

- energy is added to system; and
- Increases in the internal energy of system.
- –q and –w
- energy is removed from system; and
- Decrease in the internal energy of system.

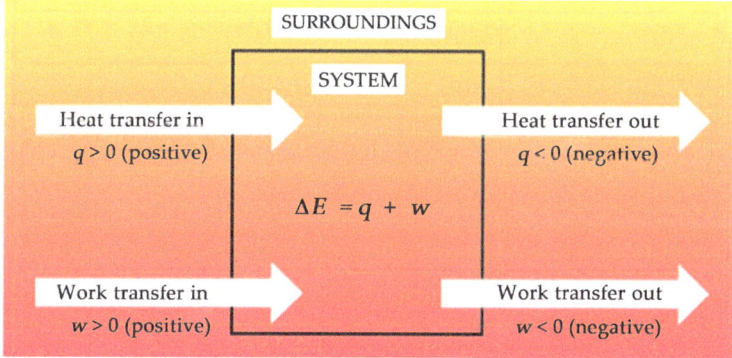

**Figure 2.1.** Sign convention of heat and work in a thermodynamic system.

## 2.1.1. Heat Transfer and Temperature Changes

When there is gain of kinetic energy (KE) by an object, then it results in a more rapid movement of the constituent atoms and molecules, while at the same time, its temperature increases. There are certainly three factors that impact the overall magnitude of the temperature change for an object: (1) the amount of heat energy added to the object, (2) the material the object, and (3) the mass of the object is made of. Let's consider lighting a match and then use it to heat a large glass full of water (H2O). There is transfer of heat from the match stick to the glass of H2O, but there is no much change in the temperature of the H2O.

Now let's consider using a lit match to heat the tip of a needle. In this case, the tip of the needle become hot so quickly. It is worth noticing that the same amount of heat energy is transferred to both objects, but the tip of the needle gets hotter comparatively to glass of H2O if both tasks were performed at the same time. This is because the needle has a smaller mass as compared to the glass of H2O, in addition is the needle being made of metal allows a more efficient flow of heat due to its conductivity, giving it a lower specific heat capacity than the glass and the H2O. Specific heat capacity refers to the amount of energy required to elevate the temperature of 1 g of a substance by 1°C (equation 2.2).

$$c, \text{specific heat capacity} (J / g.°C) = \frac{q, \text{heat energy absorbed (J)}}{m, \text{mass(g)} \cdot \Delta T, \text{change in temperature}(C°)} .........(2.2.)$$

**Table 2.1.** Specific Heat Capacity Values of Common Substances

| Substance | Name | Specific Heat (J/g· K) |
|---|---|---|
| Al(s) | aluminium | 0.897 |
| Fe(s) | iron | 0.449 |
| Cu(s) | copper | 0.385 |
| Au(s) | gold | 0.129 |
| $NH_3(g)$ | ammonia | 4.70 |
| $C_2H_5OH_{(l)}$ | ethanol | 2.44 |
| $H_2O_{(l)}$ | water | 4.186 |
| $H_2O_{(s)}$ | water | 2.06 |
| | wood | 1.8 |
| | concrete | 0.9 |
| | glass | 0.8 |
| | granite | 0.8 |
| $O_{2(g)}$ | oxygen | 0.917 |
| $N_{2(g)}$ | nitrogen | 1.04Y |

*Source: Table by Employees.oneonta.edu.*

Specific heat capacity values are basically stated in units of J/g°C or J/g·K. There is no change in the values with the different units primarily because a one-degree raise in temperature is equal on both temperature scales. Table 2.1– depicts that some materials, for instance metals, have very low specific heat capacity, which means it requires very small amount of energy to rise its temperature 1 degree higher.

With materials, like $H_2O$ having high specific heat capacities, it means that it requires higher amount of energy to rise level of temperature. For instance, the same level of heat energy will elevate the temperature of a 1-g sample of gold almost thirty times more than it would a 1-g sample of $H_2O$.

## 2.1.2. Determining Specific Heat Capacity

It is possible to determine the value of the specific heat capacity when there is particular knowledge about the changes in mass, energy, and temperature of the sample. Imagine an experiment where 5-g samples of both and a of glass are heated and 150 J of heat energy is added to both samples. The change in temperature of silver sample is far higher compared to that of the glass sample. Thus, it can be concluded that the silver has a smaller specific heat capacity compared to water.

## 2.1.3. Using Specific Heat Capacity

To work effectively with chemical and energy systems, there are some key skills required such as: calculation of the amount of energy in relation to temperature change and assessment of the magnitude of temperature change.

Example:

In the laboratory, you were asked to heat an unknown metal, weighing 55 g to 85°C and then instructed to immediately plunged it into 100 ml of water which it as at 25°C. After a few minutes, a stable temperature reading of water was obtained at 32°C. Assuming no loss of energy in the surroundings, determine the: (a) amount of energy the water absorbed; (b) the heat capacity of the metal; the (c)specific heat of the metal; and (d) explain the heat exchange mechanism of the experiment.

Given:

Mass of metal = 55 g.

T metal = 85°C

Mass of water = 100 g (assuming water density is 1 g/ml)

T water (initial) = 25°C

T water (final) = 32°C

Required:

(a)     amount of energy the water absorbed

(b)     the heat capacity of the metal

(c)     specific heat of the metal

(d)     explain the heat exchange mechanism of the experiment.

Solution:

a)     qwater =mwatercpwater (ΔTwater)

= (100 g) (4.184 J g$^{-1}$ °C$^{-1}$ ) (32 – 25) °C = 2, 928.8 J

b)     the energy absorbed by the water is equal to the energy lost by the metal. Thus, the qmetal = -2,928.8 J = Cmetal ΔTmetal; where C is the heat capacity of the metal.

$$C_{metal} = \frac{q_{metal}}{\Delta T_{metal}} = \frac{-2,928.8 J}{(85-25)^{\circ} C} = 48.81 \frac{J}{C}$$

c)     the specific heat of metal cp  from the value of qmetal = -2928.8 J will be:

$$C_{pmetal} = \frac{q_{metal}}{m_{metal} \Delta T_{metal}} = \frac{-2,928.8 J}{55g(85-25)^{\circ} C} = 0.888 \frac{J}{g^{\circ} C}$$

d).     From this experiment, as explained in b, the energy absorbed by the water is equal to the energy lost by the metal. Hot metal at 85°C will release heat to cool down as it was immersed in the water. The metal loses energy and the magnitude of such energy will be assigned a negative value since heat is lost. Simultaneously, as the metal loses heat, the water absorbs it and starts to experience rise in temperature. At a specific point in time, the metal and the water, as one system, will reach a stable temperature, called *thermal equilibrium.* In this experiment, the calculations were done based on the assumption that no heat is lost to the surroundings as the metal cools down and as the water heats up, and the 32°C is assumed to be the temperature at the thermal equilibrium.

Heat capacity by nature can be characterized an extensive quantity. This usually varies with respect to pressure and volume; thus, it does not depend on the amount of the substance because it is describing the systems as a whole. Cp is the capacity at constant pressure and Cv is at constant volume. In this problem the heat capacity is obtained in the constant pressure assumption. The specific heat (cp) on the contrary is an intensive property which can be dependent on the amount of the substance present.

## 2.1.4. Heat Transfer Between Substances: Thermal Equilibrium and Temperature Changes

When objects of varying temperatures come into contact with each other, the hotter object transmit thermal energy to the cooler object. It results in causing the hotter object to become cool, and reversely the cooler object to become warm. This process continues until both objects reach the same temperature, a state we called *thermal equilibrium*.

At any point in the transfer of heat, the quantity of heat energy lost by the hotter object (–qlost) is equivalent with that gained by the cooler object (+q gained). That is, qlost + qgained = 0. Considering now a heated iron bar plunged into H2O. In case, one defines iron as the system and similarly H2O as the surroundings, the process is exothermic as loss of heat or energy is by the system (iron) into the surroundings (H2O). The energetic changes that take place in the system and surroundings are described in Figure 2.2 as follows:

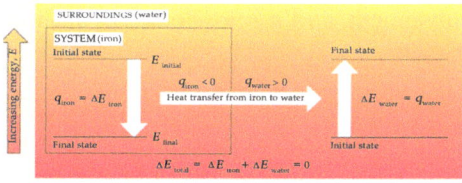

**Figure 2.2**. The dynamics of heat and energy using water and silver systems.

## 2.1.5. Energy, Changes of State, and Heating Curves

If one heat a sample of ice at approximately –10°C at 1 atm pressure, it becomes warmer. If one heat ice at 0°C (and 1 atm), however, at the initial stage, it will not become warmer. Instead, the ice will melt while sustaining a consistent temperature of 0°C, the melting point of H2O.

As it is earlier discussed, when an object is heated, it becomes warmer, or it can experience a transition phase, and it can also undergo a chemical change. In this, the changes in energy in the context of warming and cooling objects are also discussed. Now the emphasis is on the second possible outcome - which is a phase change. When a solid for example, ice is heated and melts, the heat energy is used to address the forces holding the H2O molecules together in the solid phase.

Therefore, the heat energy is changed into chemical potential energy. This potential energy can be reconverted to thermal energy when there is reformation of H2O molecules, when there is freezing of liquid water to ice.

The energy to affect a transition phase has been evaluated for a number of substances. The enthalpy  or heat of fusion (Hf) of a substance is the required energy to melt one mole or one gram of that substance while the enthalpy (heat) of vaporization or heat of vaporization (Hv) is the required energy to vaporize one mole or one gram of that substance.

## 2.2. HESS'S LAW

There are basically two methods to assess the amount of heat involved in a chemical change: measure it through experiments or estimate it from other experimentally determined enthalpy changes. There are some reactions which are difficult to do, if not impossible, to examine and make exact experimental measurements. . The estimation of the total enthalpy changes of a certain chemical system lies in obtaining some basic thermodynamic properties of the substances and a knowledge of its formation process. As stated in the Hess Law:

*"If a process can be written as the summation of numerous stepwise processes, the enthalpy changes of the total process is equivalent with the sum of the enthalpy changes of the various steps."*

Hess's law is emphasizes that because enthalpy is a state function, enthalpy changes basically depend on the preliminary stage of a chemical process, but not on the path it takes from start to finish.

Example 1:

One can take into consideration the reaction of carbon with oxygen in order to create CO2 as by a two-step process that must end up with an overall process:

$$C(s) + O_2(g) \ \square \ CO_2(g) \qquad\qquad \Delta H298° = -394 \text{ kJ} \qquad (2.3)$$

As proof of concept for Hess Law, consider the formation of carbon dioxide from the combustion of solid carbon with corresponding $\Delta H298°C$ = -394 kJ.

In the two-step process, first there is formation of carbon monoxide using equation (2.3.1) then the formation of carbon dioxide from carbon monoxide using equation (2.3.2). Add up the two equations and cancel all terms that can be seen both is the reactant and product side. As indicated, carbon monoxide from equation (2.3.1) and equation (2.3.2). Adding up the remaining terms and the enthalpy values, the overall equation goes back to equation (2.3).

$$C_{(s)} + \tfrac{1}{2} O_{2(g)} \longrightarrow \cancel{CO_{(g)}} \qquad \Delta H_{298°C} = -111 \text{ kJ} \qquad (2.3.1)$$

$$\cancel{CO_{(g)}} + \tfrac{1}{2} O_{2(g)} \longrightarrow CO_{2(g)} \qquad \Delta H_{298°C} = -283 \text{ kJ} \qquad (2.3.2)$$

$$C_{(s)} + O_{2(g)} \rightarrow CO_{2(g)} \qquad \Delta H_{298°C} = -394 \text{ kJ.}$$

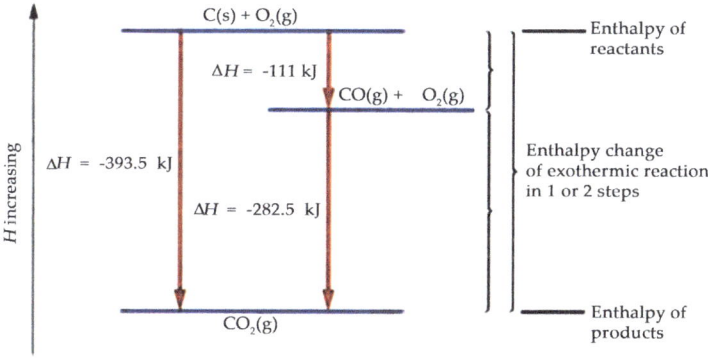

**Figure 2.3.** The 2-step formation of CO2 explained by Hess Law.

In this Figure 2.3, the creation of CO2(g) from its elements can be believed to be occurring in two steps, which sum to the overall reaction, as defined by Hess's law. The horizontal blue lines characterize enthalpies. For an exothermic process, the products are usually at lower enthalpy than are the reactants.

$\Delta H^0_{reaction} = \sum n\Delta H^0_{f(prod)} - \sum n\Delta H^0_{f(react)}$ Example 2: The standard heat of combustion of benzene is − 3271 kJ/mol. The multi-step reaction involved formation of CO2 ($\Delta H$= - 286 kJ/mol) and water at -286 kJ/mol. Calculate the standard heat of formation of benzene.

This problem requires some help for your basic General Chemistry and Basic Thermodynamics.

For the combustion of benzene, the reaction is:

$C6H6 + 15/2 \ O2 \bullet 6 \ CO2 \ + 3 \ H2O$          -3271 kJ/mol          (2.4)

Formation of CO2:    $C \ + \ O2 \ \bullet CO2$    - 394 kJ/mol          (2.4.1)

Formation of water:   $H2 \ + \ ½ \ O2 \bullet H2O$- 285 kJ/mol          (2.4.2)

To convert the 3 equations above to "formation of benzene" we need to reverse equation (2.4) which makes its ΔH positive. Also, all the 3 equations must be balanced. Please put in mind that adjustments you need to do with their corresponding ΔH, when the number f moles of the products were modified to balance it:

$6 \ CO2 \ + 3 \ H2O \bullet C6H6 + 15/2 \ O2$          -3271 kJ/mol          (2.41)

$6 \ C \ + \ 6 \ O2 \ \bullet 6 \ CO2$                    6(- 394 kJ/mol )          (2.4.1.1)

$3 \ H2 \ + \ 3/3 \ O2 \bullet 3H2O$                    3( - 285 kJ/mol)    (2.4.2.1)

Then cancel all cancellable terms in the reactants and product side, proceeding to addition with the enthalpies.

| | | |
|---|---|---|
| $\cancel{6 CO_2} + \cancel{3 H_2O} \rightarrow C_6H_6 + \cancel{15/2 O_2}$ | +3271 kJ/mol | (2.42) |
| $6 C + \cancel{6 O_2} \rightarrow \cancel{6 CO_2}$ | - 2364 kJ/mol) | (2.4.1.2) |
| $3 H_2 + \cancel{3/2 O_2} \rightarrow \cancel{3 H_2O}$ | - 858 kJ/mol) | (2.4.2.2) |
| $6 C + 3 H_2 \rightarrow C_6H_6$ | + 49 kJ | (2.5) |

Equation (2.5) is called the overall thermochemical reaction.

## 2.3. CALORIMETRY

Calorimetry is the process of measuring the amount of heat released or absorbed during a chemical reaction. By knowing the change in heat, it can be determined whether or not are action is exothermic (releases heat) or endothermic (absorbs heat).

One technique that is widely used in order to calculate the amount of heat involved in a chemical or physical process is known as calorimetry. Calorimetry is used to measure the total amounts of heat transferred to or from a substance. To do so, it is essential that the heat is exchanged with a calibrated object (calorimeter). The change in temperature of the measuring part of the calorimeter is transformed into the amount of heat (as the earlier calibration was used to know its heat capacity).

The assessment of heat transfer with the help of this approach requires the knowledge of a system (the substance or substances experiencing the physical or chemical change) and its surroundings (the other components of the measurement apparatus that play the role of either providing heat to the system or absorbing the heat from the system).

Information about the heat capacity of the surroundings, and diligent calculation of the masses of the system and surroundings as well as their temperatures before and after the process makes it possible to measure the amount of heat transferred.

A calorimeter refers to a device that is used to measure the amount of heat that is involved in a physical or chemical process. For example, when there is occurrence of an exothermic reaction in a solution inside a calorimeter, the heat which is produced in a reaction is absorbed by the solution, which surges its temperature.

When there is occurrence of an endothermic reaction, the required heat is absorbed from the thermal energy of the solution, which reduces its temperature (Figure 2.4). The change in temperature, along with the specific mass and heat of the solution, can then be used to compute the amount of heat involved in either case.

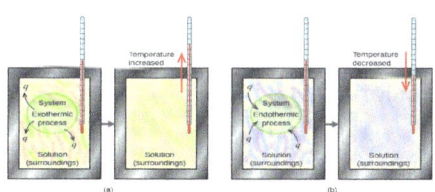

**Figure 2.1.** Calorimetry determines the amount of heat involved in a physical and chemical process.

*Source: Image by web.ung.edu.*

In a calorimetric determination, either (a) an exothermic process occurs and heat (-q),, is negative, indicating that there is transfer of thermal energy from the system to its surroundings, or (b) an endothermic process occurs and heat(+q), is positive, demonstrating that the transfer of thermal energy is from the surroundings to the system.

Scientists basically make use of well-insulated calorimeters that all but avert the transmission of heat between the calorimeter and its environment.

This result in precise assessment of the heat involved in the energy content of foods, chemical processes, and so on.

It is generally seen that the chemistry students make use of simple calorimeters made from polystyrene cups. These easy-to-use "coffee cup" calorimeters make it possible for heat to transfer with their surroundings, and therefore produce less accurate (but better than none) energy values.

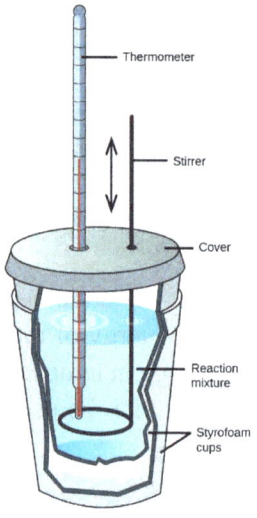

**Figure 2.5.** A simple calorimeter.

*Source: image by web.ung.edu.*

A simple calorimeter can be formed with the help of two polystyrene cups. A thermometer and stirrer spread through the cover into the reaction mixture.

There is also availability of commercial solution calorimeters. Calorimeters that are relatively cheaper consist of two thin-walled cups that are nested in a way which restrain thermal contact during use, along with a handheld stirrer, insulated cover, and simple thermometer. Calorimeters with high process tagline used for industry and research typically have a well-insulated, motorized stirring mechanism, entirely enclosed reaction vessel, and a at temperature sensor.

The calorimeters described are primarily designed to perform at a consistent level of (atmospheric) pressure and are suitable in calculating heat flow accompanying processes that arise in solution. A different type of calorimeter that also performs at a consistent volume range, is a bomb calorimeter - is used to calculate the energy released from reactions that produce large amounts of gaseous products, such as combustion reactions. (The term "bomb" arrives from the past experiences that these reactions can be dynamic enough to resemble blasts which hampers the functioning of other calorimeters.) Such kind of calorimeter consist of a robust steel container (the "bomb") that comprises the reactants and is itself submerged in H2O. The sample is put into the bomb in which oxygen is filled at high pressure. A small electrical spark is used in order to make sure that the sample catch fire.

The energy which is released in the process reaction is trapped in the steel bomb and the surrounding H2O. The increase in temperature is then measured and, along with the identified heat capacity of the calorimeter, is used to measure the produced energy by the reaction. However, bomb calorimeters need regular calibration to provide accurate heat capacity.

It is possible to accomplish calibration using a reaction with a known q, such as a measured quantity of benzoic acid exploded by a spark from a nickel fuse wire that can take place before and after the reaction. The change in temperature is the result of a known reaction that is basically used to predict the heat capacity of the calorimeter.

The calibration usually occurs each time before the use of calorimeter to accumulate the research data. Since the process is done at the same volumes, the reaction vessel must be designed in such a way to bear the high pressure which is the result of a combustion process, and is conducive to a confined explosion. The vessel is therefore referred as a "bomb," and the technique is called bomb calorimetry. The reaction is originated by discharging a capacitor with the help of a thin wire that ignites the mixture.

Another noticeable result of the constant-volume condition is that the release of heat is corresponds to qv, and therefore to the internal energy change $\Delta U$ instead of $\Delta H$. The change of enthalpy can be calculated with the help of following formula:

$$\Delta H = qv + \Delta ngRT \qquad\qquad (2.5)$$

in which $\Delta ng$ is the change in the number of moles of gases of a substance in the reaction.

## 2.3.1. How Do Calorimeters Work?

Although calorimetry is known for its easiness by principle, its practicality is highly technical art, particularly when applied to processes that occur slowly or comprised of very little heat changes, such as the germination of seeds. Calorimeters can be as simple as a foam plastic coffee cup that is mainly used in chemistry labs. Research-grade calorimeters that possess the ability to sense changes in minute temperature, are probably more likely to occupy tabletops, or even entire rooms.

The ice calorimeter is considered as an essential tool for computing the heat capacities of liquids as well as solids, in addition to the heats of certain reactions. This simple yet creative apparatus performs its role perfectly for measuring the change in volume due to melting of ice. To assess heat capacity, a warm sample is put into the inner compartment that is surrounded by a mixture H2O and of ice.

The heat released from the sample as it cools initiates some of the ice to melt. Since ice is less dense than H2O, there is a possibility of a decrease in volume of H2O in the insulated chamber. This resulted in enabling an equal amount of mercury to enter into the inner reservoir from the outside container. The decrease in container weight directs the decrease in volume of the H2O, melting the ice. This, in association with the heat of fusion of ice, gives the amount of heat lost by the sample as it cools to 0°C.

## 2.4. LAWS OF THERMOCHEMISTRY

In a thermochemical system with small variations in  entropy and energy, the molecular variations can be conceived all equal for majority of the molecules, however, in a thermochemical system with large variations in energy and entropy, the molecular variations depend on the global entropy and energy, and so on the molecule.

### 2.4.1. First Law of Thermochemistry

An isolated thermochemical system evolves irrevocably to the state of thermochemical equilibrium.

$$dU \bullet 0 \text{ as } t \bullet \infty \qquad (2.6)$$

in which U refers to the molecular energy of the isolated thermochemical system; and t represents the time.

2.4.2. Second Law of Thermochemistry

In a thermochemical system there is conservation of energy.

$$\frac{dU}{U} = \frac{dQ + dW + dE + c^2 dm + dm + d\gamma A}{Q + W + E + mc^2 + \gamma A} \qquad (2.7)$$

where U refers to the energy; W, the mechanical energy; Q, the thermic energy; mc2, the relativistic energy; E, the electromagnetic energy; and A, the surface tension of the thermochemical system.

### 2.4.3. Third Law of Thermochemistry

In a thermochemical system the sum of the variations of the molecular KE and molecular thermomechanical energy is the sum of the variations of the molecular surface tension and molecular thermal energy.

$$\frac{VdP + PdV + dK}{PV + K} = \frac{dQ + d\gamma A}{Q + \gamma A}$$

(2.8)

in which PV refers to the thermomechanical energy; Q, the thermal energy; K, the chemical KE; and A, the surface tension of the thermochemical system.

### 2.4.4. Fourth Law of Thermochemistry

In a thermochemical system the molecular KE seems to be half of the Planck's constant times the characteristic regularity of the molecular normal mode as the molecular temperature tends to zero.

$$K \rightarrow \frac{1}{2}\hbar v \text{ as } T \rightarrow 0$$

(2.9)

in which K refers to the molecular KE; T, the molecular temperature of the thermochemical system; and the Planck's constant, the characteristic occurrence of the molecular normal mode.

### 2.4.5. Fifth Law of Thermochemistry

In a thermochemical system, the difference of the molecular entropy is the difference of the molecular thermal energy less the molecular relativistic energy per molecular temperature.

$$\frac{dS}{S} = \frac{dQ - \sum a_{ik}\mu_{ik}d\xi_k}{(T_m / T)(Q - mc^2)}$$

(2.10)

in which S refers to the entropy; Q, is known as thermic energy; T, the global temperature; Tm, the molecular temperature; mc2, the relativistic energy; and a$ik$ and $ik$, the stoichiometric coefficients and chemical potentials of the $i$ chemical reactants and products in the k reversible and irreversible energy exchanges and irreversible chemical reactions with extent of k reaction of the thermochemical system.

## 2.4.6. Sixth Law of Thermochemistry

In a thermochemical system, the difference of the molecular entropy is primarily because of the irreversible variations of the molecular KE and molecular thermomechanical energy and there are always positive irreversible chemical reactions.

$$dS_i \geq 0 \qquad (2.11)$$

in which Si refers to the molecular entropy of the thermochemical system because of the unalterable thermochemical processes.

*Seventh Law of Thermochemistry.* In a thermochemical system, it is more likely that the molecular entropy is zero as the molecular temperature seems to be zero.

$$S \to 0 \text{ as } T \to 0 \qquad (2.12)$$

in which S refers to the molecular entropy; and T is called the molecular temperature of the thermochemical system.

## 2.4.7. The van der Waals Thermochemical Equation

A differential behavior of the van der Waals Equation on pressure in liquids and solids arises, as greater differentiation on pressure give small differences in quantity, and similarly, small differences in the intermolecular forces of collision, which suggests that the van der Waals Thermochemical Equation for solids and liquids is:

$$Pe^{aN^2/V^2}(V - Nb) + K = Q + \gamma A \qquad (2.13)$$

where P V is the thermomechanical energy; K refers to the chemical KE; Q, is called the thermal energy; A, is the surface tension; and N, is the total molecular number; and a and b, the coefficients of van der Waals of the thermochemical system.

## 2.4.8. The Earth Thermochemical Equation

By the Third Law of Thermochemistry, as the atmosphere is considered as a thermochemical system with no fluctuating heat capacity within a spherical ring volume with a radius H for which small differentiation of temperature arise within hefty fluctuations in pressure, slight variations of chemical KE, and little variations of radius, its thermochemical differential equation is

$$\frac{H_0 dP + P dH + dK}{H_0 P + K_0} = \frac{dT}{T_0} \qquad (2.15)$$

and so,

$$(H_0P + K_0)e^{(HP+K)/(H_0P+K_0)} = (H_0P_0 + K_0)e^{T/T_0}$$ (2.15.1)

where K refers to the chemical KE; P, is the pressure; T, the level of temperature; H, is called the radius of atmosphere; K0, refers to the chemical KE, whereas P0, the pressure; T0, the temperature; H0, is the radius of the atmosphere at sea level.

By the Third Law of Thermochemistry, as the Earth is a thermochemical system with constant heat capacity in which no fluctuation of surface tension and little variations of temperature arise within large variations of radius, large variations of pressure, and small variations of chemical KE, its thermochemical deferential equation is

$$\frac{HdP+PdH+dK}{HP+K_0} = \frac{dT}{T_0+\gamma A_0}$$ (2.15.2)

and so, equation 2.15.3 is:

$$(HP + K_0)e^{K/(HP+K_0)+T_0/(T_0+\gamma A_0)} = (H_0P_0 + K_0)e^{T/(T_0+\gamma A_0)+K_0/(HP+K_0)}$$

where K refers to the chemical KE; P, is the total pressure; T, is denoted as the temperature; H, is called the radius of the Earth; K0, is the chemical KE; whereas P0, is the pressure; T0, is the temperature; H0, denoted as the radius; and A0, is the surface tension of the Earth at sea level.

By the Third Law of Thermochemistry, as the nucleus of Earth is conceived as a thermochemical system with changeable heat capacity in which greater alterations of thermal energy and surface tension arises within greater alterations of pressure, small alterations of radius and greater alterations of chemical KE, its thermochemical deferential equation is

$$\frac{\frac{4}{3}\pi H_0^3 dP+4\pi H^2 PdH+dK}{\frac{4}{3}\pi H_0^3 P+K} = \frac{dQ+d\gamma A}{Q+\gamma A}$$ (2.15.4)

and so,

$$(Q_0 + \gamma A_0)(\tfrac{4}{3}\pi H_0^3 P + K)e^{\tfrac{4}{3}\pi H^3 P/(\tfrac{4}{3}\pi H_0^3 P+K)} = (\tfrac{4}{3}\pi H_0^3 P_0 + K_0)(Q + \gamma A)e^{\tfrac{4}{3}\pi H_0^3 P/(\tfrac{4}{3}\pi H_0^3 P+K)}$$ (2.15.5)

where K refers to the chemical KE; Q, is the thermic energy; P, is the pressure; H, the radius; and A, the surface tension of the magmatic core; K0, is known as the chemical KE; Q0, is called the thermic energy; P0, is the overall pressure; H0, the radius; and finally, A0, is known as the surface tension of the magmatic core surface.

## 2.5. SYSTEMS AND PROCESSES

System is the matter that is under observation or in consideration—it gives information on the total amount of products and reactants in a chemical reaction. It can be the amount of solute and solvent used to form a solution. It can be the gas inside a balloon.

Then, the environment or surroundings refers to everything which is outside that system. The system is being separated from the surroundings by a boundary. However, the boundary between surroundings and system is not permanently fixed, and thus can be shifted. For example, the mass of coffee in a coffee cup can be considered as a system and the cup holding it refers to the environment . This kind of setup is useful for someone that is interested in knowing the total amount of heat transferred from the hot coffee to the cooler coffee cup.

Otherwise, one might consider the system as the hot coffee as well as the cup together, and the air surrounding the coffee cup as the environment. This type of setup is useful in calculating the heat exchange between the hot coffee and cup system and the cooler surrounding air. The boundary can be extended as farther depending on the circumstances until the whole mass of the universe is eventually counted in the system. In this case, there will be no surroundings. Again, if one placed a boundary, it is primarily based on what phenomenon one is interested in studying.

Systems is basically considered by whether one want to exchange matter or heat with the surroundings. It can be summarized as follows:

Isolated: In this the system cannot exchange energy (work as well as heat) or matter with the surroundings; for instance, an insulated bomb calorimeter.Closed: In this the system can exchange energy (work as well as heat) but not matter with the surroundings; for instance, a steam radiator. Open: In open system, it is possible to exchange both energy (work and heat) as well as the matter with the surroundings; for instance, a pot of boiling $H2O$.When a system acknowledges a transition phase in one or more of its properties (such as concentrations of products, or reactants, pressure or temperature), it undergoes a process.While processes, by meaning, are events interconnected with a change of the state of a system, some processes are exclusively identified by some property that is continuous throughout the process. Majority of these processes form special conditions because it allows to simplify the first law of thermodynamics:

$$\Delta U = Q - W$$
(2.16)

where ΔU refers to the alteration in internal energy of the system, Q is the total heat sum up to a system, and W is the total work done by the system.

For example, isothermal processes occur when the system's temperature is constant. Constant temperature implies that the total internal energy of the system (U) is constant throughout the process. This is because temperature and internal energy are directly proportional. When U is constant, ΔU = 0 and the first law simplifies to Q = W (the heat added to the system equals the work done by the system). An isothermal process appears as a hyperbolic curve on a pressure–volume graph (P–V graph). Work is represented by the area under such curve, as shown in Figure 2.6.

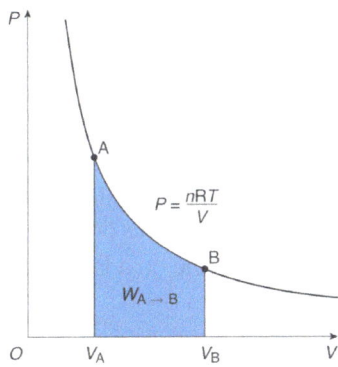

**Figure 2.6.** Graph of an isothermal expansion temperature.

*Source: Image by schoolbag.info.*

The graph of an Isothermal Expansion Temperature is constant in an isothermal process; thus, the area under the curve represents not only the work performed by the gas, but also the heat that entered the system.

Adiabatic processes occur when no heat is exchanged between the system and the environment; thus, the thermal energy of the system is constant throughout the process. When Q = 0, the first law simplifies to ΔU = –W (the change in internal energy of the system is equal to work done on the system [the opposite of work done by the system]). An adiabatic process also appears hyperbolic on a P–V graph, as shown in Figure 2.7.

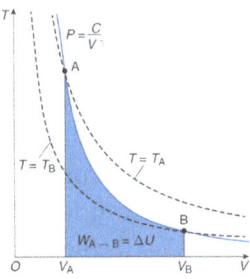

**Figure 2.7.** Graph of an adiabatic expansion.

*Source: Image by schoolbag.info.*

Heat exchange is zero in an adiabatic process; temperature is not constant (as shown by the dotted lines).

Isobaric processes occur when the pressure of the system is constant. Isothermal and isobaric processes are common because it is usually easy to control temperature and pressure. Isobaric processes do not alter the first law, but note that an isobaric process appears as a flat line on a P–V graph, as shown in Figure 2.8.

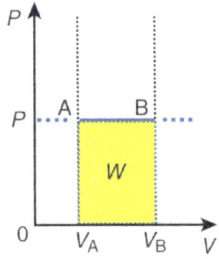

**Figure 2.8.** Graph of an isobaric expansion.

*Source: Image by schooobag.info.*

Pressure is constant in an isobaric process; the slope of the line is therefore zero.

Finally, isovolumetric (isochoric) processes experience no change in volume. Because the gas neither expands nor compresses, no work is performed in such a process. Thus, the first law simplifies to $\Delta U = Q$ (the change in internal energy is equal to the heat added to the system). An isovolumetric process is a vertical line on a P–V graph; the area under the curve, which represents the work done by the gas, is zero.

## 2.6. STATE FUNCTION DEFINITION

Imagine that you want to climb Mount Snowdon in Wales. Wales is the little country just left of England. You can get up to the top of the mountain in two different ways. You can spend two to four hours hiking to the 3,560 feet peak or you can take the train, which is a nice, leisurely 1-hour ride to the top. Because you can't hike up in a straight line, your hiking distance is much more than the height of the mountain. The train, on the other hand, has a much more direct route to the peak.

Both ways will get you to the top, and in both cases, your altitude from start to finish will change by 3,560 feet. Critically, it didn't matter how you got to the top of the mountain, the change in altitude was the same. This is an example of a state function, which is a property whose value does not depend on the path taken to reach that specific value.

In other words, it didn't matter whether you climbed the mountain by foot or went by train, you still ended up 3,560 feet higher up than when you started. Your final position in both cases was the same. Now, not everything is a state function. Let us explore the way you got up the mountain. If you hiked up, you put in a lot more effort to do the climb.

But if one goes by train, then you put in very little effort to reach to the mountain as it is the train that is doing all the work for a person. So, the amount of effort, undertaken by a person, was different and primarily rely on the route chosen by a person to reach up the mountain. This effort undertakes by a person to reach his destination is called work.

Work is a path function and not a state function, because the value varies on the route chosen by the person. One question that is relevant in figuring out whether something is a state function is: is it important to know how one arrive at a final value do? In case the answer is 'no,' then it is a state function.

State functions are "variables" which play the role of defining the state of a system. When a person has a system, then it is essential to define the conditions in which it exists before and after a change. It is basically referred as the initial and final states.

By states it means the system refers to a set of properties. For example, the state of a system may contain a 1 mole of argon in a 10 L container with a temperature of 300 K. Here the state of the system is defined by the "state functions" of temperature and volume as well as the quantity of the gas. In a similar way, the pressure is also a state function. Later, in thermodynamics, it is also observed that there is a numerous variable in relation to energy that are also state functions.

If 1 mole of gas is taken to do something, there might be a case that one end up in a new state where for instance the quantity is around 20 L and the temperature is 600 K. For this kind of process, there is a need to take into consideration the change in the state functions. In this scenario, the initial volume Vi, was 10 L and the final volume, Vf, was 20 L. So, the change in volume is given by

$\Delta V = Vf - Vi = 20L - 10L = 10L$

The primary objective to this belief is that the change in volume has no relation with the particulars on the mystery process that cause to move from the initial state to the final state. The variance in volume will always be the same. That is, for instance, if one starts at a volume of 10 L and ended with a volume of 20 L the difference will always remain same which is +10L. This might seem frightfully obvious.

However, if one take into notice about abstract state functions in relation with energy, then it could be more difficult to wrap around the ideas. But it is essential to recognize that the concept is exactly the same. If one has understanding about the initial state as well as the final state, then it is important to calculate the change (regardless of the process by which the change is achieved).

## 2.7. ENTHALPY

Enthalpy (H) is defined as the sum of the internal energy of a system plus the product of pressure as well as volume.

The formula is:

$$H = E + PV \tag{2.17}$$

It has been observed that in most chemical systems under study, reactions are carried out under conditions of constant pressure. Under this condition, the change in enthalpy, that is $\Delta H$, is equal to the heat that is exchanged under constant pressure. It has to be noted that, enthalpy cannot be measured directly like the internal energy of a system, and also it is not possible to know the amount of enthalpy that is present in a chemical sample. However, enthalpy change and thus, relative enthalpy, can be evaluated.

Enthalpy measures the total heat content of a system and it is related to both chemical potential energy as well as the degree to which all the electrons are attracted to the nuclei in molecules. In the case when the electrons are strongly attracted to the nuclei, there are strong bonds between atoms and molecules are relatively stable, then enthalpy is low.

On the other hand, when electrons are only weakly attracted to the nuclei, there are weak bonds between atoms. The molecules are relatively unstable, and enthalpy is high in this case. The sign of H shows the direction of energy transfer. Heat is transferred from the system to the surroundings in an exothermic reaction. For an exothermic reaction, the enthalpy change has a negative value (that is H < 0).

The molecules which are weakly bonded are converted to strongly bonded molecules during exothermic reactions. In addition to it, chemical potential energy is transformed into heat, and there is an increase in the temperature of the surroundings. On the other hand, in an endothermic reaction, heat is transferred from the surroundings to the system. For an endothermic reaction, the enthalpy change has a positive value (that is H > 0).

Strongly bonded molecules are converted to weakly bonded molecules during endothermic reactions. In addition to it, heat is converted into chemical potential energy, and there is the decrease in the temperature of the surroundings.

## 2.7.1. Representing Energy Change

The physical and chemical changes a substance undergoes in the process has been established to be accompanied by the changes in enthalpy. These changes on the other hand is according to how they utilize heat in the process. The heat utilization of a chemical reaction as discussed in the earlier sections can be exothermic or endothermic. Figure 2.9 illustrates how heat or energy utilization behavior according to the type of reaction. In these diagrams, the horizontal axis shows the transformation of the system which is basically from the reactants state to product formation. The vertical axis, on the other hand indicates the relative enthalpy of every single state.

There is an increase in enthalpy as one moves up the vertical axis. In an endothermic reaction, the reactants require a much higher energy than that of the product. This means that the initial state has higher energy requirement than that of the final state. This is the reason why, the change in enthalpy of an exothermic reaction is characterized to be negative ($-\Delta H$ = Hfinal – Hinitial). Conversely, this process is opposite when the reaction is endothermic. The positive change in enthalpy is due to the fact that the energy of the products is higher than that of the reactants.

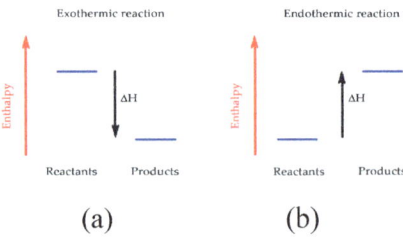

(a)                    (b)

**Figure 2.9.** Schematics of energy trending in an (a)exothermic and (b) endothermic chemical reaction.

## 2.8. ENERGY, TEMPERATURE CHANGES AND CHANGES OF STATE

There are three (3) things can happen when an object is heated. When the object (1) gets warmer, (2) it can go through a phase change, and (3) can also undergo a chemical change.

## 2.9. HEAT (Q)

Heat is a measurable form of energy and it is indicated by changes in temperature. When you bring 2 bodies of different temperature, heat flows from the hotter to the colder body. In most cases (but not all), when heat is transferred, the hotter object cools down and its temperature lowers and the coder body heat up of its temperature increases. Temperature (T), therefore, is also associated to the average KE of the particles in a substance.

In a substance, the average KE of the particles is associated to the thermal energy (that is enthalpy) of the substance. But because there is the inclusion of consideration of how much substance is present in order to evaluate total thermal energy content, the most that can be said about temperature is that when thermal energy of the substance increases, its temperature also increases.

Nevertheless, it cannot be said that something that is hot necessarily has greater thermal energy (in absolute terms) as compared to the other substance that is cold. For instance, it can be determined that a large amount of lukewarm H2O has a greater total content of heat as compared to a very small quantity of hot H2O.

## 2.9.1. Key Concept of Heat

It has to be noted that heat and temperature are two different things. Heat is defined as a specific form of energy that can enter or leave a system. On the other hand, temperature is a measure of the average KE of the particles that are present in a system.

Kelvin is the absolute temperature scale. It was determined through the third law of thermodynamics that there is a finite limit to temperature below which nothing can exist. It means that there will be no molecular activity below 0 K. This is because by the definition, the system is said to be unable to lose any more heat energy at this point. However, quantum mechanics explains about a state of molecular motions that is possible below absolute zero.

Recalling that heat (Q) is the transfer of energy from one substance to the other substance, this transfer takes place as a result of a difference in temperature. The zeroth law of thermodynamics states that when the temperature of the objects is equal, objects are in thermal equilibrium.

Therefore, heat is defined as a process function, it is not a state function. It has been observed that one can measure the amount of thermal energy that is transferred between two or more than two objects as a result of their difference in temperatures by calculating the heat that is transferred.

On the other hand, the first law of thermodynamics explains that the change in the total internal energy ($\Delta U$) of a system is equivalent to the amount of heat (Q) that is transferred to the system excluding the amount of work (W) that is performed by the system.

The formula (from equation 2.16) is:

$\Delta U = Q - W$.

One can measure the transfer of energy in the form of heat through any kind of process irrespective of the amount of work that is done. This is because heat and work are calculated independently. Endothermic ($\Delta Q > 0$) are the Processes in which the system absorbs heat. On the other hand, those kind of processes in which the system releases heat are known as exothermic ($\Delta Q < 0$).

The unit of heat is equivalent to the unit of energy: joule (J) or calorie (cal), for which 4.184 J = 1 cal. Enthalpy ($\Delta H$) is equal to heat (Q) under constant pressure.

One of the most essential ways that the body works in order to prevent overheating is by the production of sweat. Sweat is an exocrine secretion of electrolytes, H2O, and urea. Though, it is not the production of sweat that is the cooling mechanism. It is defined as the evaporation of the sweat. This helps in cooling the body Evaporation (that is vaporization) from the liquid to gas phase is an endothermic process.

Energy must be absorbed from the body for the particles of the liquid to get enough quantity of KE. This energy is used to escape into the gas phase. It has been observed that hot, arid desert air has a lower partial pressure of H2O vapor as compared to the humid, tropical air. Therefore, sweat evaporates more readily in the dry air as compared to that in the humid air. So, it has been observed that most people will feel more comfortable in dry heat as compared to that in humid heat.

When substances of different temperatures are brought into thermal contact with each other—that is, some physical arrangement that permits the transfer of heat—the flow of energy will be from the warmer substance to the cooler substance. On the other hand, when a substance goes through an endothermic or exothermic reaction, heat energy will be exchanged between the system as well as the environment.

Calorimetry is defined as the process of measuring the heat that is transferred. There are two basic types of calorimetry. This consists of constant-pressure calorimetry and constant-volume calorimetry. For instance, the coffee-cup calorimeter, is a low-tech example of a constant-pressure calorimeter. On the other hand, a bomb calorimeter is an example of a constant-volume calorimeter.

The heat (q) that is absorbed or released in a given procedure is measured by the equation:

$q = mc\Delta T$

It is also characterized as a form of energy that is linked with the random motion of the elementary particles that are present in matter. Heat can be classified whether it is:

Heat capacity – The amount of heat that is required to increase the temperature of a defined amount of a pure substance by one degree.

Specific heat – The amount of heat that is required to increase the temperature of one gram of a substance by 1°C (or 1 K)

Calorie – The specific heat of H2O = 4.184 J/g-K

SI unit for specific heat is joules per gram–1 Kelvin–1 (J/g-K)

Molar heat capacity – The amount of heat that is needed to increase the temperature of one mole of a substance by 1°C (or 1 K)

The SI unit for molar heat capacity is joules per mole-1 Kelvin–1 (J/mol-K)

Btu (British thermal unit): The amount of heat that is required to increase the temperature of 1 lb. H2O by 1°F.

**NOTE:** The specific heat of H2O (4.184 J/g-K) is very large as compared to the other substances. The oceans (that cover more than 70% of the earth) act as a giant "heat sink."

These oceans are moderating drastic changes in temperature. The temperatures of the body are also controlled by H2O and its high specific heat. Perspiration is defined as a form of evaporative cooling which keeps the temperatures of the body from getting too high.

Latent Heat versus Sensible Heat

Sensible heat: Heat which can be easily spotted with a change in the temperature of a system.

Latent heat: Heat which is not easy to detect because the temperature of the system exists at the same level.

- e.g., The heat which is used to evaporate H2O or to melt ice is latent heat.

There are basically two categories of latent heat:

- Heat of fusion – The heat which should be absorbed in order to melt a mole of a solid.
- e.g., melting ice to liquid H2O
- Heat of vaporization – The heat which should be absorbed to boil a mole of a liquid.
- e.g., boiling liquid H2O to steam

Caloric Theory of Heat

- Served as the basis of thermodynamics.
- Now considered as an obsolete
- Based on certain assumptions discussed as follows:
    - Heat is a fluid which means it passes from hot to cold substances;
    - Heat usually attracts more strongly to matter which can hold a lot of heat;

-     Heat is conserved;
-     Sensible heat mostly intensifies the temperature of an object when it started flowing into an object;
-     Latent heat associates with particles in matter (initiating substances to boil or melt); and
-     Heat does not have any weight.

Only one thing that is valid about the caloric theory is that heat is weightless.

Heat is NOT a fluid, as it is NOT conserved.

Kinetic Theory of Heat

- •     Splits the universe into two categories:
1.     System – The small portion of the universe in which one particular have an interest.
2.     Surroundings – In the system, everything is not included, for instance, the rest of the universe.
- •     A boundary separates both the system as well as surroundings from each other and can be imaginary or tangible.
- •     Heat is something which is transported back and forth across boundary between system and surroundings around it.
- •     Heat is not conserved.
- •     The kinetic theory of heat majorly relies on the last postulate in the kinetic molecular theory (KMT) which propose that the average KE carrying group of gas particles is primarily rely upon the temperature of the gas.

$$KE_{avg} = \frac{3}{2}RTn$$

(2.18)

where R refers to the ideal gas constant (0.0821 L-atm/mol-K), and T is temperature (Kelvin).

## 2.10. CONCLUSION

There are various interrelated terms which is important in understanding the concepts of thermochemistry and distinguishing it from thermodynamics and applied chemistry. With the interrelations previous discussed in this section, heat and energy are one of the main factors that communicates these three disciplines.

Energy is the capacity to perform the work at certain levels (applying a force to move matter). KE refers to the energy of motion; potential energy is energy due to composition, relative position, or condition. When there is conversion of energy from one form into another, it is accompanied by the transfer of heat.

Specific heat and heat capacity help in estimating the energy required to change the temperature of an object or substance. The amount of heat released or absorbed by a substance depends directly on the type of substance, its weight, and the temperature change it undergoes.

Calorimetry also play very important role as it helps in measuring the amount of thermal energy transferred in a physical or chemical process. This requires cautious measurement of the temperature variation that arises during the process and the masses of the system and surroundings. These measured quantities then help in computing the amount of heat consumed or produced in the process using known mathematical relations.

# REFERENCES

1.  Ch301.cm.utexas.edu. (n.d.). State Functions. [online] Available at: https://ch301.cm.utexas.edu/section2.php?target=thermo/properties/state-functions.html (accessed on 13 March 2020).

2.  Chapter 5: Thermochemistry, (2020). [ebook] Available at: http://employees.oneonta.edu/viningwj/Chem111/Chapter_05_4_SY.pdf (accessed on 13 March 2020).

3.  Chem1.com. (n.d.). Thermochemistry and Calorimetry. [online] Available at: http://www.chem1.com/acad/webtext/energetics/CE-4.html#SEC4 (accessed on 13 March 2020).

4.  Chem.libretexts.org. (2020). Calorimetry – Chemistry LibreTexts. [online] Available at: https://chem.libretexts.org/Bookshelves/Physical_and_Theoretical_Chemistry_Textbook_Maps/Supplemental_Modules_(Physical_and_Theoretical_Chemistry)/Thermodynamics/Calorimetry (accessed on 13 March 2020).

5.  Chem.tamu.edu. (n.d.). Thermodynamics: Temperature, Heat, and Work. [online] Available at: https://www.chem.tamu.edu/class/fyp/stone/tutorialnotefiles/thermo/thermo.htm (accessed on 13 March 2020).

6.  Chemistry LibreTexts, (n.d.). 6.4: Quantifying Heat and Work. [online] Available at: https://chem.libretexts.org/Bookshelves/General_Chemistry/Map%3A_A_Molecular_Approach_(Tro)/06%3A_Thermochemistry/6.03%3A_Quantifying_Heat_and_Work (accessed on 13 March 2020).

7.  Chemteam.info. (2020). Energy, Heat, Work, Temperature. [online] Available at: https://www.chemteam.info/Thermochem/Energy-Work-Heat-Temp.html (accessed on 13 March 2020).

8.  Cordero G. D., (n.d.). The Laws of Thermochemistry. [ebook] Available at: https://vixra.org/pdf/1407.0154vD.pdf (accessed on 13 March 2020).

9.  Khan Academy, (2020). Heat and temperature (article) | Khan Academy. [online] Available at: https://www.khanacademy.org/science/chemistry/thermodynamics-chemistry/internal-energy-sal/a/heat (accessed on 13 March 2020).

10. Schoolbag.info. (2020). Systems and Processes – Thermochemistry – Training MCAT General Chemistry Review. [online] Available at: https://schoolbag.info/chemistry/mcat_2/49.html (accessed on 13

March 2020).

11. Schoolbag.info. (n.d.). Heat – Thermochemistry – Training MCAT General Chemistry Review. [online] Available at: https://schoolbag.info/chemistry/mcat_2/51.html (accessed on 13 March 2020).

12. Sciencedirect.com. (2016). Calorimetry – An Overview | ScienceDirect Topics. [online] Available at: https://www.sciencedirect.com/topics/chemical-engineering/calorimetry (accessed on 13 March 2020).

13. Study.com. (2020). State Functions in Thermochemistry – Video & Lesson Transcript | Study.com. [online] Available at: https://study.com/academy/lesson/state-functions-in-thermochemistry.html (accessed on 13 March 2020).

14. Thermochemistry, (2012). [ebook] S Energy Information Administration. Available at: https://web.ung.edu/media/chemistry/Chapter5/Chapter5-Thermochemistry.pdf (accessed on 13 March 2020).

15. Www2.southeastern.edu. (2020). Thermodynamics Part 1: Work, Heat, Internal Energy and Enthalpy. [online] Available at: https://www2.southeastern.edu/Academics/Faculty/wparkinson/help/thermochemistry/ (accessed on 13 March 2020).

# Structure/Reactivity and Thermochemistry of Ions

## CONTENTS

This chapter is about the development and new theories about the thermochemistry of ions. In this chapter, different bonds formation, structure, and their working mechanisms are explained. Such as, ionic bonds, covalent bonds and reactions like Kossel–Lewis theories. This section also explained how ionic bonds are formed in ionic solids, energy consideration in ionic structure, and how thermodynamics is used in the ion exchange process, equilibrium and in isotherm.

## 3.1. INTRODUCTION

Matter is said to be made up of one or different types of elements. Except noble gases, there are no other element that exists as an independent atom in nature, under normal conditions. There are groups of atoms that are found to exist together as a one species that are said to have characteristics properties. Such groups of atoms are called a molecule. While to form a molecule, there is some force that must hold the constituent atoms together.

The force that hold these atoms, ions and particle together to form a molecule is called a chemical bond.

The result of combination of the same or different atoms in different ways are new sets of substances exhibiting new sets of properties based on the type of chemical bond that holds them . There is certain question that arises due to this reason. Like, why do atoms combine? Why are only certain combinations possible? Why do some atoms combine while certain others do not? Why do molecules possess definite shapes? So, to answer these questions there are certain different theories and concepts that were put forward from time to time.

Some of the theories are Kossel-Lewis approach, Valence Bond (VB) Theory, Valence Shell Electron Pair Repulsion (VSEPR) Theory and Molecular Orbital (MO)

Theory. The nature of the chemical compound and the evolution of these theories on valences are considered to be closely related with each other. It helps in understanding the development of the structure of the atoms, the electronic configuration of the element and the periodic table. Fundamentally, atoms combine to be more stable.

The combination of atoms to form stable orbitals requires the transfer or sharing of electrons, which must be undertaken with adequate amount of force or chemical energy.

## 3.2. IONS

"Anion is an atom or a molecule with the net electric charge, due to loss or gain of one or more electrons."

Anions are negatively charged atoms and cations are the positively charged atoms.

An ion is an atom or molecule with a net electric charge due to the loss or gain of one or more electrons. There are specialized types of ions commonly known as anions and cations. Anions are considered to contain a greater number of electrons than protons. So, anions are considered to contain net negative charge.

Cations are positively charged ions because it has more protons than electrons...

Anions are considered to be generally larger than the parent molecule or atom. This is so, because of the excess of electrons get repelled by each other. It also adds to the physical size of the electron cloud. The cations are generally considered to have smaller size then the parent atom or molecule. This is because of the smaller size of their electron clouds.

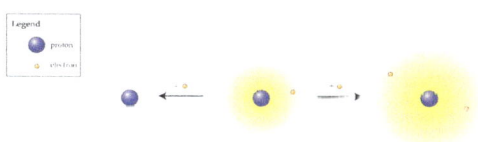

**Figure 3.1.** Formation of cations and anions.

*Source: Image by Wikimedia Commons.*

**Table 3.1.** Comparison between Molecules, Compounds, Cations and Anions

| Cations | Anions | Compound | Molecule |
|---------|--------|----------|----------|
| $Na^+$ | $H^-$ | HCl | $H_2, N_2, O_2$, etc. |
| $Mg^{+2}$ | $Cl^-$ | $CH_4$ | $P_4$ |
| $Al^{+3}$ | $O^{-2}$ | $CaCl_2$ | $CaCl_2$ |
| $Fe^{+2}, Fe^{+3}$ | $S^{-2}$ | $H_2SO_4$ | $H_2SO_4$ |

Ions are written as chemical or molecular formulas with superscripts carrying positive and negative notations. Table 3.1 above illustrates the comparison between anions (cations and anions), compounds and molecules. Cations and anions as defined carries changers with it. Their charges are written as superscripts of their chemical symbols. Monoatomic ions are also sometime represented by roman numerical. This designates the formal oxidation state of the element. On the other hand, the superscripted numerals are denoted by the net charge. For example, Fe+2 can also be referred to as Fe (II) and Fe+3 can also be referred to as Fe (III). However, Roman numerals cannot be applied to a polyatomic ion. Polyatomic ions, sometimes called radicals in chemistry, are group of atoms carrying charges. A classic example of a cationic radical is ammonia, NH3+. On the other hand, sulfate (SO4-2), and nitrate (NO3-1) are classified as negative radicals.

On the other hand, one must delineate the compound and molecules based on the types of atoms present. All compounds are molecules but not all molecules are compounds. Molecules are theoretically defined as a mixture of 2 or more like or unlike atoms or compounds, while compounds are combination of 2 or more unlike substances. All substances written under the compound column in Table 3.1 are also molecules. They are under compound because they are made up of 2 or more "unlike" substances. While, H2 and other diatomic substances are not compounds, they are molecules, because still they are a combination of 2 or more "like" substances.

> Monoatomic is an anion consisting of exactly one atom. If anion contains more than one atom, even if these are of the same element, it is called a polyatomic ion.

## 3.3. FORMATION OF IONS

Ionization is one of the processes used for forming ions. Ionization is a process where neutral atoms loose or gain electrons. In general, the electrons are either added or removed from the valence shell of an atom. The inner shell of the atom is said to have electrons that are more tightly bound to the positively charged nucleus. So, in this way, the nucleus do not participate in this type of chemical interaction.

Ionization is said to generally involve the process which involves the transfer of electrons between atoms and molecules. This process is more motivated, so as to achieve a more stable electronic configuration. One that gives atoms a stable configuration is by obeying the octet rule, which states that most of the atoms and ions that are stable, are said to have eight electrons in the outer most shell or valence. Generally, by the process of gaining and losing the elemental ions, such as H+ ions in the neutral molecules, polyatomic, and molecular ions are said to be formed. These polyatomic ions are generally said to be very unstable and reactive. However, the octet rule is not absolute for all atoms since there are atoms that are exempted from satisfying the octet rule.

Common example of a cation is Na+ or sodium ion. Sodium ion is said to have a +1 charge because it has eleven electrons. However, according to the octet rule, with 10 electrons, i.e., 2 in the inner most shell and 8 in the outermost shell, sodium is said to be more stable. Therefore, because of this reason sodium tends to lose its 1 outer electron to another atom become more stable, than accepting 7 electrons more. The energy required to transfer this single to electron of sodium to another atom is called its ionization energy. There is obviously lesser energy needed to transfer 1 electron than transferring 7.

On the other hand, chlorine has already 7 electrons in its outermost shell and accepting 1 more electron will make it stable based on the octet rule. This makes chlorine ion as Cl-.

## 3.4. TYPES OF BONDS

### 3.4.1. Ionic Bond

Ionic bond is also called electrovalent bond. It is a type of linkage that are formed from the electrostatic attraction that takes place between the oppositely charged ions in a chemical compound. Such bonds are formed when the valence or the outermost shell electrons of one atom is transferred permanently to another atom. The cations are the atoms that loses the electrons and becomes positively charged ions. While the anions are the ones that gains the electrons and become negatively charged ions.

When complete transfer electrons take place from one atom to another, then the ionic bond is said to be formed. This is takes place so that both of the atoms acquire stable electronic configuration. When the elements of group 1 and 2 combines with the halogens, oxygen, and sulphur, ionic bonds are formed. Similarly, when a metallic substance combined with a non-metallic substance, ionic bond is formed.

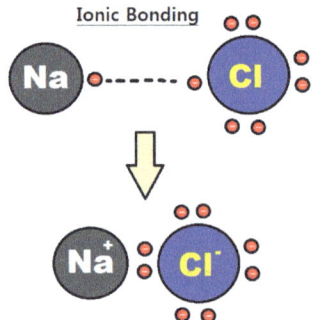

**Figure 3.2.** Representation of ionic bond.

*Source: Image by Wikimedia Commons.*

As ionic bonds can be best exemplified by the compounds that are formed between non-metals and the alkali and the alkaline earth metals, the transfer of electron from sodium atom to the chlorine atom forms an ionic crystalline solid. This, NaCl is a substance held by the electrostatic forces of attraction between opposite charges and repulsion between similar charges. In addition, the strong electrostatic forces allows ionic compound to form organized crystalline lattice as illustrated in Figure 3.2.1.

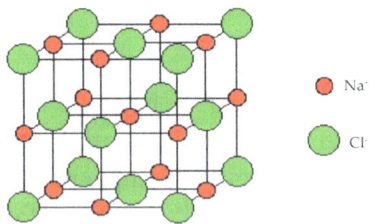

**Figure 3.2.1.** The organized crustal structure of an ionic compound NaCl.

In short it can be explained that the ions are arranged in such a manner that the positive and negative charges alternate and balances one another with the overall charges of the entire substance equal zero. In case of ionic crystals, the magnitude of the electrostatic forces is considerable. Accordingly, these substances are tended to be hard and non-volatile.

There are several conditions that are necessary for the formation of an ionic bond:

    1.    Atoms forming positive (cation) ions (usually metals) should

have:

- Low electron affinity;
- Low ionization Energy;
- Low electro negativity; and
- High lattice Energy

2.  Atoms forming negative (anions) ions (usually non-metals) should have

- High electro negativity;
- High ionization energy;
- High electron affinity; and
- High lattice energy.

An ionic or electrovalent bond has the following characteristics:

- When there is an attraction between the positively and negatively charged ion, an ionic bond is said to be formed.
- An ionic bond is non-directional, i.e., the strength of interaction between two ions depend upon distance, but not on the direction.
- The actual transfer of the electrons from the metal to the non-metallic atoms makes the bond non-directional, making bond breaking energy extensive.
- An ionic bond gets broken when the substance is dissolved in a polar solvent or when the substance is melted.

## 3.4.2. Covalent Bond

When the atoms of the same or different elements combines by mutual sharing of electrons then a covalent bond is formed. The compound which is formed is known as a covalent compound. These chemical bonds are formed by electrons sharing orbitals. Eexample of a compound that is formed by covalent bonding is an $O_2$ molecule, which follows that all diatomic substances are covalently bonded.

Electrons pairs of the covalently bonded atoms spins though the 2 atoms overlapped orbitals.. When 2 atoms identical electronegativities combine, their outer orbitals overlap forming a symmetric electron cloud

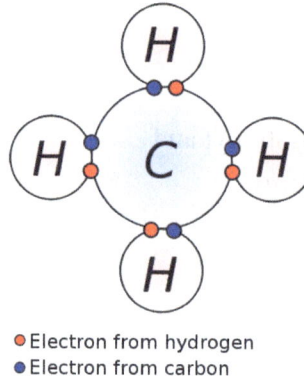

Electron from hydrogen
Electron from carbon

**Figure 3.3**. Representation of covalent bonds in hydrogen and carbon.

*Source: Image by Wikipedia.*

The term "Covalent bond" first came into use around 1939. Although the term "covalence" was first introduced by Irving Langmuir in the year 1919. This covalence is described by the number of electron pairs that are shared by the neighbouring atoms.

Bonding pairs or shared pairs are known as the electrons pairs that participates in formation of a covalent bond. Typically, these bonding pairs sharing allows each atom to achieve stable outer electron shell configuration. Similar case is seen among noble gases.

There are several conditions that are necessary for the formation of covalent bonds:

- A bond formed by mutual sharing of electrons;
- Formed between two or more non-metals –
- A single bond represents two electrons;
- They satisfy the octet rule by sharing of electrons; and
- These molecules have a definite shape.

### 3.4.3. Polar and Nonpolar Covalent Bonds

Covalently bonded molecules can be further classified based on the electronegativities of combining substances. With respect to diatomic substances, the combination is by two substances having identical electronegativities. When their orbitals overlap to facilitate electron sharing, the symmetrical electron cloud indicates non-polarity of the substance. As such, when the atoms share electron pairs equally then the non-polar bonds occur. Another is to look at the difference is the electronegativities of the combining substance. When the difference exceeds 0.5 as indicated by the Pauling's scale, the compound or molecule has a non-polar covalent bond. Diatomic substances have a zero difference in their electronegativities. Similarly,  CH4 is classified to have non-polar bonds because of a 0.4 difference in the electronegativities.

With the increase in the electronegativity difference, the electron pairs in a bond is more closely associated with the one nucleus. This happens when two substances of different electronegativities combine. In the case of HCl, we know that it exhibits covalent bonding since they are both non=metallic by nature. Looking into the difference in their electronegativities gives us an idea that it exhibits polar covalent bond because of a 0.9 difference. With this, hydrogen which is more electronegative (3.0) behaves as the partially negative atom and chlorine which is less electronegative (2.1) behaves as the partially positive atom. When they combine, they form an asymmetric electron cloud. Additional note must be considered that, if the electronegativity difference is lower than 0.5, it is has non-polar covalent bond; when it lies between 0.5 and 1.7, then this bond is said to be polar covalent bond, and  the electronegativity difference of greater than 1.7is a characteristic if an ionic compound/molecule.

### 3.4.4. The Ionic Lattice

A giant structure of ions is known as ionic compounds. If these ions arc having a regular and repeating arrangement, then this is called an ionic lattice. The formation of lattice takes place because the ions attracts each other and forms a regular pattern with the oppositely charged ions next to each other. An example of an ionic lattice is illustrated in Figure 3.4.

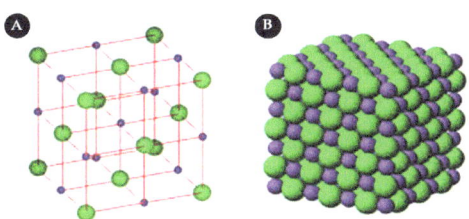

**Figure 3.4.** An example of an ionic crystalline lattice formed through ionic bonding.

## 3.5. REPRESENTATION OF IONIC COMPOUNDS

There are different types of model available that can be used to represent the giant ionic structure. Each giant structure has its own advantages and limitations. For examples:

- The Two-Dimensional space-filling model: This model clearly shows the arrangement of ions. This arrangement takes place in one layer, but it does not show how the next layer of ions is arranged.

- The Three-Dimensional ball and stick model: This type of model shows the arrangement of ions in a larger section of the crystal. But using sticks for bonds is misleading, as the forces of attraction between ions actually act in all directions.

- The Three-Dimensional Model: This model is also misleading because it shows lots of free space between the ions, which is generally not present.

### 3.5.1. Kossel -Lewis Approach to Chemical Bonding and Formal Charge

To represent the valence electron in an atom, G. N. Lewis, an American chemist, introduced simple notations commonly known as Lewis Symbols. The study of the electronic configuration of noble gases is being studied by Lewis and Kossel. They also observed that the inertness of the noble gases is due to their complete octet or duplet. This is in the case of Helium which contains 2 electrons in its last shell. They also emphasized that, "the atoms of different elements combine with each other in order to complete their octets or duplets (in case of H, Li, and Be) to attain stable electronic configuration." Thus, they generalized that some atoms might not obey octet rule but still attains a stable configuration. The Lewis Dot Symbol or sometime called the Lewis Dot structure illustrates bonding electrons and electron pairs using dots. The structure is then finalized by representing bonding electron pairs by line and unpaired electrons stays as dots in the structure.

## 3.6. REPRESENTATION OF STRUCTURE

Lewis structure: In this case, it represents the valence electrons of the atom in the form of dots. With the help of Lewis structure one can easily recognize if any of the atoms possess lone pair electrons or have a formal charge. For example, $H_2O$ in Figure 3.5.1, and hydroxyl radical and in Figure 3.5.2

hydroxide ions can be represented with the help of Lewis structure as shown below:

| (a) | (b) | (c) |
|---|---|---|
| H· ·Ö· ·H | H:Ö:H | H–Ö–H |

**Figure 3.5.1.** The correct Lewis Dot Structure of water (H2O) is in letter (c).

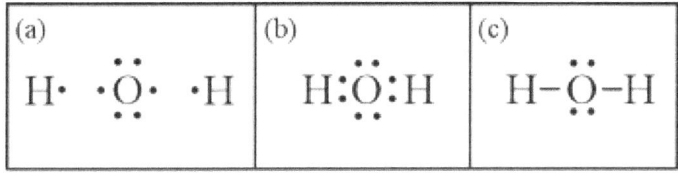

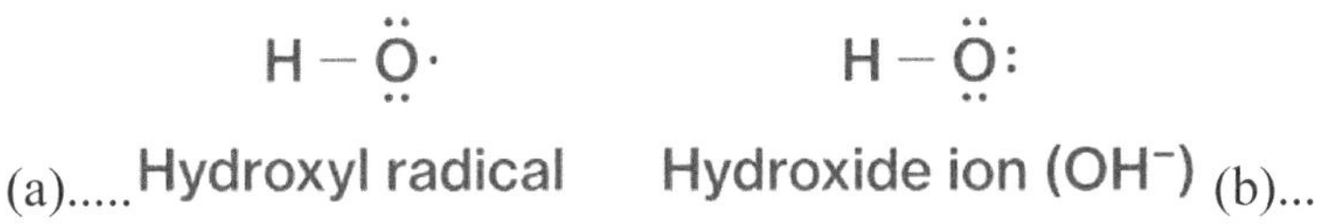

(a).....Hydroxyl radical       Hydroxide ion (OH⁻) (b)...

**Figure 3.5.2.** The correct Lewis Dot Structure of hydroxyl radical (a) and (b) hydroxide ion.

The valence electrons that are not used for bonding, such electrons are called non-bonding electrons or lone-pair electrons, which are retained in the Lewis Dot Structure. When a positive or negative charge is assigned to compound or a molecule, then this is called formal charge. In the case of H2O there is no formal charge, but the oxygen atom in the hydroxide ion is considered to have a formal charge of –1. It is important to construct first the Lewis Dot structure of the compound or molecule to determine correctly its formal charge. The formal charge of a compound or molecule is a 2-step process. Calculate first the formal charges of each of the atomic components, then determine the formal charge of the molecule by simply adding up all the formal charges of the atomic component of the molecule. The formula for the formal charge of each component can be written as:

$$\text{formal charge} = \text{number of valence electrons} - \left( \text{number of lone-pair electrons} + \frac{\text{number of bonding electrons}}{2} \right)$$

(3.1)

The structure of ammonia, ammonium ion and amide anion are shown below using Lewis structure. In this structure Nitrogen contain five valence electrons. Ammonium ion is said to contain formal charge of +1 and amide anion on the other side contain formal charge of –1.

```
                    H
                    ··+
H : N̈ : H      H : N : H        H : N̈ : H
    H                H               ··
  ammonia      ammonium         amide anion
                  ion
```

By employing equation 3.1., the formal charges of ammonia, ammonium ion and amide ion is:

**it is best to remember that all Hydrogens here had a formal charge of zero $[ 1-(0+1)] = 0$

Nitrogen in Ammonia:      $5 - [1 + (6/2)] = +1$

      in Ammonium ion: $5 - [0 + (8/2)] = +1$

      in Amide ion:      $5 - [2 + (8/2)] = -1$

Formal Charges:

  Ammonia:      $N + 3(H) = 1+(3x0) = +1$

  Ammonium ion: $N + 4(H) = 1 + (4x0) = + 1$

  Amide ion:  $N + 2(H) = -1 + (2x0) = - 1$

Carbon is said to contain four valence electrons. The structure of methane, methyl cation, methyl anion and methyl radical using Lewis structure is shown below:

```
     H               H               H               H
     ··              ··              ··              ··
H : C : H       H : C +         H : C : ⁻        H : C ·
     ··                              ··              
     H               H               H               H
  methane      methyl cation    methyl anion    methyl radical
```

When a species contains a positively charged carbon atoms then, that is known as carbocation. But when a species containing negatively charged carbon atom then, that is known as carbanion. When a species containing an atom that contains a single unpaired electron, this is known as free radical. Hydrogen contains only one valance electron,  while halogens like F, CL, Br, I, contains seven valence electrons. The indication of formal charges of the species like hydride ion, hydrogen radical, bromine, bromine radical, bromide, have been given below:

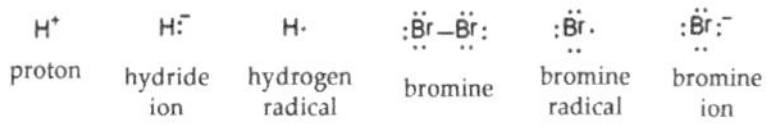

H⁺          H:⁻          H·          :B̈r—B̈r:          :B̈r·          :B̈r:⁻
proton    hydride     hydrogen    bromine      bromine      bromine
            ion         radical                  radical        ion

The lines between two atoms can also show the pair of shared electrons. The hydrogen atom consists of one covalent bond and said to have no lone pairs. Whereas, in the case of hyalogens, it has one covalent bond and consists three lone pairs. H2O molecule consist of oxygen which has two covalent bond and also two lone pairs. In case of ammonium atom, the nitrogen bond consists of three covalent bonds and also one lone pair. Each of these atoms are said to has a complete octet except the hydrogen atom which is consider to be completely filled with the outer shell.

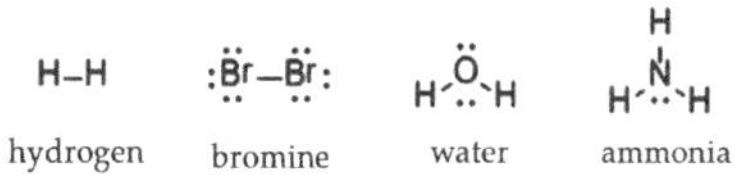

H–H          :B̈r—B̈r:          H–Ö–H          H–N̈–H
                                                        (with H above)
hydrogen     bromine          water            ammonia

Kekulé structures: In this type, the structure of the atom is represented by the bonding electrons as a line and the lone pairs electrons are left out entirely. This happens unless they are needed to be drawn a mechanism for a chemical reaction.

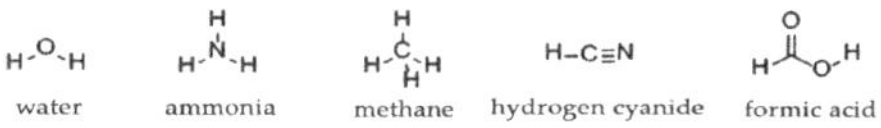

H–O–H     H–N–H     H–C–H     H–C≡N     H–C(=O)–O–H
water      ammonia    methane    hydrogen cyanide   formic acid

Sometimes, there are some of the covalent bonds of the structure of a compound that can be omitted for the simplification. These types of structure are known as condensed structures.

$H_2O$          $NH_3$          $CH_4$          HCN          HCOOH
water          ammonia        methane      hydrogen cyanide   formic acid

## 3.6.1. The Lewis Structures of a Polyatomic Ion

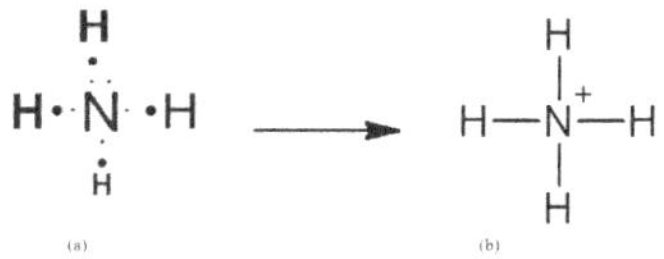

H
·
H··N··H     ⟶     H—N⁺—H (with H above and below)
·
H

(a)                              (b)

**Figure 3.6.** Lewis structure of ammonium; (a) electron sharing by pairing, (b) actual structure with valence/charge.

*Source: Image by Socratic.*

The valence or charges of polyatomic ions or radicals is determined similarly as the determination of the formal charge of a molecule. On the other hand, the Lewis Dot structure is very helpful in doing so because it gives you an idea of the number of shared electrons as well as the lone pairs.

Another method of determining the valence or charge of a radical is by visual inspection of its structure and some basic knowledge of periodic properties of the atomic components of the radical. Looking at Figure 3.6, Ammonium [NH4]+ has a positive charge, which was proven as calculated in section 3.6 of this chapter. This again can be done by inspecting the molecular structure of the radical:

1.  Based on the data available for Nitrogen, it has 5 valence electrons (it belongs to Group 5A). Writing fundamental Lewis Dot structure mean it must have 5 dots around it.

2.  It is common knowledge now that Hydrogen can only have 1 electron for sharing.

3.  Pairing the 1 electron of hydrogen to that of nitrogen, leaves one electron unpaired (see Figure 3.6 a). Adding 1 more hydrogen to pair with the excess will violate the molecular formula and it will no longer be ammonium in doing so.

4   The remaining unpaired electron is now the one that makes the radical more active, indicating that once, it engages in a chemical reaction, then unpaired electrons will be the first one to be transferred or shared. As such, ammonium has a tendency to remove one electron to be stable.

5   Since it is required to remove 1 electron for ammonium to become stable, the final structure is written by replacing paired electrons with lines and that unpaired electron to be removed will leave nitrogen a positive charge. Looking at Figure 3.6.b, we know that it is only N which will have charges since the hydrogens are stably bonded, giving the whole radical a charge of +1.

**Table 3.2.** This Table Represents the Bonds of Neutral Covalent Compounds and the Polyatomic Ions in the Different Families Like Carbon, Oxygen, Hydrogen or halogen and nitrogen.

| Family | Bonds of neutral covalent compound | Bonds in polyatomic ions |
|---|---|---|
| Carbon Family | Bonds four times | Bonds four times |

| Oxygen family and Beryllium | Bonds twice | Bonds once or twice |
|---|---|---|
| Boron or Nitrogen family | Bonds three times | Bonds three or four times |
| Halogens or Hydrogen | Bonds once | Bonds once |

*Source: Table by Chemfiesta.org.*

# 3.7. ION-EXCHANGE REACTION

When chemical reactions take place, new substances are formed. These new substances, theoretically are formed by the exchanges of ion pairs participating as reactants in a chemical reaction. Basic chemistry tells us the different types of chemical reactions that may take place. It is noteworthy to mention that carbon dioxide can be formed by combustion or combination or direct union with carbon and oxygen. Opposite of direct union reactions are decomposition reactions which happened by the action of heat. Thus, when you heat calcium carbonate, it decomposes into calcium oxide and carbon dioxide.

On the other hand, another type of reactions takes place, which is generally called substitution which entrains exchanges of ion pairs. Substitution can be single of double, but the ion exchange mechanism is more illustrated in a double substitution reaction.

When ionic substances dissolves in H2O, then their ions are set to be unconstrained, to a considerable extent from the restraints that holds them within the rigid array of the crystal. They also said to move about in the solution with relative freedom.

There are certain insoluble materials that bears the positive or negative charges on their surface to react with the ionic solutions. This happens because to remove the various ions selectively and replace the ions with other kinds. Such process is known as the ion exchange reaction.

On the other hand, not only ionic solutes are capable of forming ionic solutions. Acids and bases upon contact with water dissociates into a pool of positive and negative ions inside the solution.

The stream of the positive and negative ions makes the solutions electrolytic.

The conservative ions exchange mechanism can be observed between two compounds. He simple neutralization reaction – a reaction between an acid and a base, illustrates the efficient ion exchange mechanism that gives way to the formation of a salt (an ionic solute) and water.

Ion exchange reactions is used to remove ions from the solution in a various way and also to separate the various kinds from one another. These types of separations are widely utilized in the scientific laboratories that may affect the purification and aids in the analysis of unknown mixtures.

On the higher end, ion exchange materials such as zeolites which are also employed commercially to purify H2O and also used to medically serve as artificial kidneys and for purposes.

## 3.7.1. Early History

The idea of ion exchange has been conceived in as early as the 18th century. In 1850, agriculturalist HSM Thomson and chemist J.T. Way introduced the word "base exchange" in their separate papers published in the *Journal of the Royal Agricultural Society of England.*

The term was used to describe the substitution of potassium K+ soils from fertilizers by equal amounts of Mg2+ and Ca2+ by the presence of rainwater. This made them believe that substances tend to exchange ion pairs when subjected to conducive experimental conditions. The term "base exchange" persisted until 1940, is now termed as ion-exchange as its mechanism was understood over the years.    Nine years later, Swedish chemist Svante Arrhenius introduced the ionic theory, accompanied by his theory on acids and bases and reactions governing such phenomenon. By 1903, several synthetic organic ion exchangers were already identified and are made up of aluminosilicates.

In 1944, Gaetano F. D'Alelio patented divinylbenzene polymers as ion exchange resins. These are usually substances containing large, network-like molecules into which ionic groups are introduced by chemical treatment. This invention has introduced the use of ionic cross-linking groups such as sulfonic and quaternary ammonium groups acting as ion exchange resins.

In 1955 the ion-exchange science has been intensified by the use of phosphates, arsenates and molybdates of Titanium, Zirconium and thorium as cation exchangers. The use of molecular sieves followed which are also derivatives of crystalline aluminosilicates.

To date, the use of ion exchange agents is very important in the environmental, medical and industrial fields. The use of membranes and media containing definite-sized pores where specific-sized ions can enter was also realized, as well as selective membranes that absorbs gases.

## 3.8. IONIZATION ENERGY

The ionization energy (IE), sometimes called (less correctly) the ionization potential (usually designated by IE, IP, or, in the older literature, I), is the energy required to remove an electron from a molecule or atom:

$$M \rightarrow M+ + e- \qquad \Delta Hrxn= IEa \qquad (3.2)$$

Ionization energies are characterized as adiabatic or vertical values.

### 3.8.1. Ion-Exchange Procedures

Using ion-exchange, the bio-molecules are purified using different separating techniques that separate them according to the differences in their specific properties. In this process, the use of resin for the purpose of ion exchange is a common method. In this process the resin is mixed with the solution in a container and then replaced with the other ions of the same electrical charge. The resin itself is composed of organic polymers that helps in the formation of hydrocarbons. Throughout the polymer matrix, there are ion exchange sites where "functional groups" either positively charged ion (cations) or can be negatively charged ions (anions) are fixed to the polymer networks. These functional groups are said to attract the ions of an opposing charge.

There is another method that is used for exchanging the ions and that is using chromatography. The technique uses the process of biomolecule purification technique for ion separation. In this technique, the exchanger is placed into a tube or column and the solution is made to flow through it. The column is arranged in such a way that it forces the ion exchange reaction take place, which is considered to be intrinsically reversible and allows the completion of reaction in a controlled desired manner. The solution which is flowing down the column, continuously meets the fresh exchanger. The reaction that goes half the way in the first centimetre of the column many be three quarters completed in the second centimetres and so on.

In this short time, the ions that were exchanged enters the column, gets absorbed and becomes undetectable analytically. These exchangeable ions start to emerge from the end of the column. Therefore, the columns become completely saturated with them. There is a possibility that they be restored to their original condition or can be regenerated by passing through another solution of ions.

The use of ion exchange columns is considered to be easy. But the theory behind their use is extremely complicated. A column that contain a solution flowing through it is considered as a non-equilibrium system. With this, its interpretation must not only consider the equilibrium distributions that takes place but also the rate of transfer of material and its statistical variations in the flow path of liquid between the granules.

There are two chief theories of ion exchange processes in which the columns are working. In the first procedure, the displacement takes place when the column originally contains mobile ions that are of one kind. Then these ions are pushed down the column by the use of a steady flow solution of the ions of a second kind.

The theory for the above procedure deals with the rate at which the "front" or the boundary that is between the different classes of ions advances. It also deals with its concentration profile that whether the front stays sharp when it moves down the column or becomes progressively more diffusible.

In the second process of elution, A thin layer of ions is introduced at the top of the column that is already saturated with ions of a second kind. After this, it washes down the column with the same kind of ions. Then these ions saturate the column at the start. In the theory of elution, the rate of movement of the narrow band of the ions must be taken into account. After then it flows as it proceeds down the column.

When a single solution that contains two or more kinds of ions are held by the exchanger with the different strengths are introduced at the top of a column, these mixture of ions separates as it moves down the column. As a result, this single mixture of solution gives two or more separate bands, according to the availability of the types of ions in the solution. This process is called ion exchange chromatography.

Ion exchange chromatography plays an important tool in the field of chemical analysis because it permits the separation of materials that are found difficult to separate by other means. It can also be applied to the organic and inorganic solutions as well as to substances that are not ionic. It is often used to separate the mixture comprising complex components.

The columns that are used for ion exchange are made in all the sizes ranging from 0.5 cm to 2 m. In case of large reactors, they are used to soften hard $H2O$ supply in big cities that are holding less than a cubic centimetre of the resin that are used for recovering short lived radioactive elements in the laboratory just like the discovery of the element mendelevium by the isolation of a few atoms on an ion exchange column.

While preparing for columns for the laboratory use, extra care is needed. Before inserting resins into the column, resins must be first mixed with $H2O$.

This process will help the resin to swell and will allow easier ion exchange process. Before putting the resins, one must make sure that no air bubbles will form in the resin bed since it will interfere with the liquid flow.

It is necessary to backwash the column because it allows the liquid to pass upwards and expand the resin bed. It is also used to release the air bubbles and segregate the resin particles according to their sizes. The aim behind the preparation procedure is to assure the right packing and achieve correct flow profile.

Resins having uniform beads are considered mostly for difficult chromatographic separations. In this case very small particles are used to achieve mass transfer and also gives sharp bands during the process. The solution is forced with the high pressure since fine particles offers much resistance to flow. One of the procedures that is used for a long basis is, use of narrow stainless-steel column. These are used in case of related process caked during gas chromatography. To slow down the mass transfer, glass spheres that are coated with ion exchange resins are used instead of resin beads.

In case of thin layer chromatography, ion exchangers that are especially inorganic and cellulose based are used. By using finely ground resins and cellulose fibres, chromatographic papers for the thin layer chromatography are specifically manufactured. One of the uses for using this procedure is to filter small traces of metal ions that is from the large volume of solutions.

Ion exchange resins can also be used to make the thin sheets using fabrication process. It might not be easy to make sheets of ion exchanger that is flexible, strong, and at the same time it is permeable since there is slow development in this field of membrane technology. One of the uses of this thin sheets of ion exchange membrane is to separate the electrodes of the fuel cells. It is also used to physically remove the salts from the H2O by the process known as reverse osmosis and electrodialysis. .

Another method of separation where ion exchange is used is the kind of filtration process that squeezes H2O under pressure through the membrane. During this process the salt dissolved are left behind. This reaction can also be explained by a membrane of cations exchange resins that is placed with the sodium ions.

This process is done in contact with the dilute solution of NaCl. The entrance of the Na and Cl ions in membrane becomes restricted and allows only the passage of water.

As H2O molecules penetrate the membrane because of the pressure exerted on the system, the salts remain behind. As a result, the salts were separated from - H2O. This process is famously called today as desalting or desalination.

Electrodialysis (ED) is another process that employs the use of ion exchange membranes. This is an electrically driven membrane process that uses alternating cationic and anionic membrane to separate solutions. Often used in the treatment of salt and brackish water, the cations pass through the cationic membranes leaving the anions behind, while the anionic membrane only allows the passage of anions leaving the cations behind. This process starts, when voltage is applies. The passing through of the cations and anions through the membrane is due to the electrical current that draws cations towards the anode and anions towards the anode (see Figure 3.7).

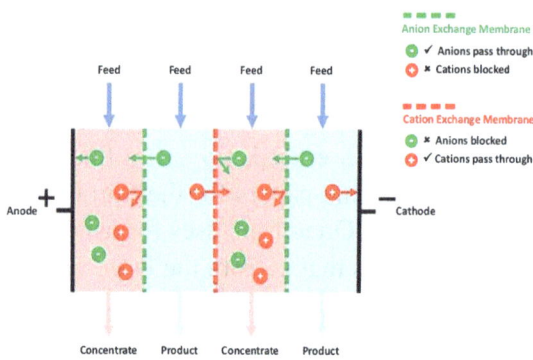

**Figure 3.7.** The Electrodialysis Process

This mechanism of ions exchange results in 2 flow regimes as end products. One is an enriched stream, or the concentrate stream, which contains the left behind cations and anions and the clean stream, or the diluate stream which are ion free.

The main difference of ED over other ion exchange membranes is that this process uses several cationic and anionic membranes in alternating positions, which they call the cell pairs. An ED stack can have as much as 200 cell pairs. In addition, this process will not work unless a potential difference is applied, meaning ions will not flow without current (Mendoza et al., 2014;2019).

### 3.8.2. Adiabatic Ionization Energy

The lowest energy that is required to affect the removal of an electron from the molecule or atom is known as the adiabatic ionization process. This corresponds to the transition that takes place from lower electronic, vibrational, and rotational levels of the isolated molecule. To the lowest electronic, vibrational, and rotational level of the isolated ions. In order to obtain the value of formation of M+ ions, adiabatic IE data is used.

### 3.8.3. Vertical Ionization Energy

The energy change that corresponds to an ionization reaction which leads to the formation of the ion in the configuration, is known as vertical IE. Its configuration is same as that of the equilibrium geometry of the ground state of neutral molecule. According to the Frank-Condon principle, when a molecule gets ionized by the process of photo ionization or by the interaction that happens with energetic electrons, then the higher probability configuration of the resulting ions will be the configuration of the precursor neutral molecules. When the equilibrium geometry of the ions become similar to that of the neutral precursor molecule, then the energies like vertical or adiabatic ionization energies will become nearby same or so. The vertical IE should be always greater than or equal to the adiabatic IE.

## 3.9. APPEARANCE ENERGY

The experiments in which the ionization photon or electron energy varies until its appearance is observed is their threshold measurements, these experiments help often in determining the ionization energies. These ionization energies is sometimes known as the appearance energies.

Therefore, this terms in general, has come with the more specific meaning. In most of the technical literature and the way it is used here, this term appearance energy sometimes can also be called the appearance potential. This refers to the minimum energy that is required to form a particular fragment ion from the precursor neutral molecule.

### 3.9.1. Proton Affinity and Gas Phase Basicity

The stable cations that are formed in the gas phase also includes the protonated neutral molecule. These neutral molecules are generated in the proton transfer reactions process. Officially, the relationship between the enthalpy of the formation of MH+ and the neutral counterparts, that is M, can be defined by the terms of the quantity known as proto affinity (PA). This negativity of the enthalpy change of the hypothetical protonation can be explained by the following reaction:

M + H+ → MH+

$\Delta$Hrxn = −PA

$\Delta$fH;(MH+) = $\Delta$fH°(M) + $\Delta$fH

(H+) − PA

The proton affinity is the term that is universally used. It is the quantity that can be defined at a finite temperature-generally 298K. Because of this, it is not strictly analogous to the adiabatic IE or electron affinity (EA). Both of these reactions have the value of K as 0, that is the enthalpy change of the corresponding reaction. At the temperature 298 K, the enthalpy of the formation of the proton is 365.7 kcal/mol, 1530.0 kJ/mol using ion convention or stationary electron convention.

The Gibbs energy change gets associated with the protonation reaction, which is called the gas basicity (GB) of molecule M. As most of the available data on the gas phase basicity and the proton affinities are being obtained from the experiment in which the equilibrium constant of the proton gets transfer reaction that is being determined.

## 3.9.2. Electron Affinity (EA)

The EA of the molecule for anions can be defined as the quantity that is analogous to the IE to have positive ions. That means, that the EA is equal to the energy difference that is between the enthalpy formation of a neutral species and the enthalpy formation of the negative ions of the same structure. The EA can also be defined as the negative of the 0 K enthalpy change, that is for the electron attachment reaction:

M + e− → M−

$\Delta$Hrxn = −EAa

-EA = $\Delta$fH

(M−) − $\Delta$fH°(M) − $\Delta$fH;(e−)

When the numeric value of the vertical quantity is being greater than or equal to the adiabatic value, then it is possible to have either vertical or adiabatic EA, that is same with the IE. The major difference from the IE is that for the stable or bound negative ion bound, the ion is said to have the lower energy than the corresponding neutral state. If in any case, negative ions are higher than the neutral ions, then the neutral ion is said to have a negative EA. This ion will usually undergo with the spontaneous loss of the electrons.

### 3.9.3. Enthalpies of Formation of Ions in the Gas Phase

The enthalpy of formation or the heat formation of the ion in the gas phase can be also obtained through the straightforward treatment of the thermochemistry of the ionization process. For example, for a positive molecular ion, the enthalpy of formation can be obtained by simply adding the enthalpy of the formation of the precursor molecules. This is done by the adiabatic IE and also by subtracting the enthalpy of the formation of the electrons. The reaction be following:

$$\Delta fH;(M+) = \Delta fH°(M) + IEa - \Delta fH;(e-)$$

Meanwhile the anion is based on the analogous quantities that can be combined with the value for the EA:

$$\Delta fH;(M-) = \Delta fH°(M) - EA + \Delta fH;(e-)$$

Similarly, the enthalpies of the formation of positive fragment ions, that is A+, and can be given as:

$$\Delta fH;(A+) = \Delta fH°(AB) - \Delta fH°(B) - \Delta fH;(e-) + PA$$

For this it should be assumed that there are no potential barriers in the reaction coordinates, that is for the deformation reaction and for little or no kinctic shifts.

It should be noted that the equation given here is with no reference to temperature. But during the practice one must be concerned about the fact that the IE or EA is the quantity that is corresponding to a 0K process. Meanwhile, the available data on the enthalpy of formation of the neutral species may correspond to some of the higher temperatures.

Because of this reason, this presents a complication in the derivations of enthalpies, that takes place during the formation of ions in the gas phase. The treatment of the enthalpies of formation of ions, that takes place at the finite temperatures requires a consideration of the changes in the IE or enthalpy of formation with temperature.

Therefore, when high accuracy is not required, the simplifying assumptions that is adiabatic IE, is approximately equivalent to the 298 K enthalpy of ionization, which is usually said to be adequate.

# 3.10. THERMOCHEMICAL CONVENTIONS FOR EN-THALPIES OF FORMATION OF IONS

Using ionization or appearance energies data, the derivation of enthalpy of formation of negative ions and positive ions in the gas phase takes place. Which is necessary for the treatment of thermochemistry of the electrons. There are basically two conventions, both of them are used widespread. They are also used for dealing with the thermochemistry of the electrons. The ion convention is used by the ion chemistry or physics community. The electron convention is used by the thermodynamics people.

## 3.10.1. The Ion Convention

The ion convention is sometime can be called the stationary electron convention. The ion convention can be explained as the enthalpy of formation of the electron at non-zero temperature which is equal to the integrated heat capacity of the electron. When data is treated using ion convention process, the enthalpy of formation of the electron cancels out the calculation of the enthalpy of formation of an ion. This means that the convention process, which is adopted by most of the mass spectrometrist and the other ion physicists and chemists, when deriving an ion enthalpy of formation, the thermochemistry of the electron can be ignored.

## 3.10.2. The Electron Convention

To treat the electron as a standard chemical element with an enthalpy of formation, that is zero at all temperature, Thermodynamicists commonly use the electron convention process. Since the standard thermodynamics is considered as a non-rationalized system. The effect that is constraining the enthalpy of the formation of the electron gas to zero in all the temperature, is the electrons are integrated in heat capacity. They are named in the derived values for the enthalpy of formation of the ions.

In case of standard thermodynamics, the data compilations and the values that are cited for the enthalpies of formation of ions at the temperature that differs from 0 K from those given before. The standard thermodynamics data compilations and the integrated heat capacity of an electron gas, may have been commonly taken to be at the same as an ideal gas following Boltzmann statics.

The value of ion convention enthalpy is 5/2 RT, the formation of a positive ion which is numerically more negatively that is, the value in the electron convention which is by 5/2 RT (6.2 kJ/mol at 298 K), by the same amount is more positive than for negative ions.

# 3.11. ELECTRONS IN MOLECULES

## 3.11.1. Ions in Ionic solids

Electron can be transferred from one neutral atom which is uncharged to another atom. During this process the atom from which this electron is transferred becomes a positive ion, and atom to which the electron is transferred becomes a negatively charged ion.

This transfer of an electron to or from the atom involves high amounts of energy. The energy which is required by the atom to remove the electron is based on the characteristics of the atom and also based on the IEs. If the atom gains the electron, energy is released by an atom to which it attracted (EA).

This electron transfer will only take place with the release of energy of the atom. This happens when the EA of the atom receiving electron is greater than the IE of the donating atom. This happens in very rare cases, in the case the resulting ions that gets associated by themselves into a new configuration is of lower energy. Most of the electrons transfer reactions occurs when the ions once formed associated themselves into structured lattice which is called an ionic crystal.

## 3.11.2. Energy Considerations in Ionic Structures

The IE of the atom that is required by the atom to produce any cation from the atom is said to be larger. The energy required for the release of electron should be larger than the energy that is required for the addition of the electron of any atom. This energy which is required for the addition of electron is known as EA. This is said to be true even if an electron is being removed from an atom which loses it, relatively easily: example is sodium and when this is being added to the atom which readily acquires it, like chlorine.

The IE of Na is +496 kJ/mole, while the EA of chlorine is only −349 kJ/mole. The reaction

$$Na(g) + Cl(g) \longrightarrow Na+(g) + Cl-(g)$$

would require +147 kJ/mole to proceed. While the reaction

$$Na(s) + 1/2\ Cl2(g) \longrightarrow Na+(g) + Cl-(g)$$

requires +377 kJ/mole. The EA alone cannot provide sufficient energy to form ions or ionic structures. Therefore, the energy must be from the assembly of isolated ions into stable multi-ion structures.

The ions that have the same type of charges are said to repel each other. But in case the ion is of opposite charges; they are said to attract each other. The simplest ionic structure that is possible, which can be stable in the gas phase ion pairs is consist of one cation and one anion, that is held together by the electrostatic attractions. For calculations, it is easy to calculate how much energy would be regained by this association by using coulomb law of electrostatic attraction. The energy of attraction can be given by:

$$E = (2.31\ x\ 10{-}16\ \text{J-pm})\ Z{+}Z{-}/d$$

In the above equation, Z is the charge on the cation and anion. Distance between the ions in pm is given by d. The energy of the two associated ions will be considered is less than the energy of the two isolated ions only if the ions are of opposite charges. In case of Na ion, the ionic radius is of 97 pm and for the Cl ion is 181 pm. So, the distance between the separation of the centres of these two atoms will be 278 pm. When the energy for one ion pair is multiplied by the Avogadro number NA, this will give the molar energy of ([Na+Cl-](g)) relative to the molar energy of the isolated ions, which can be written as:

$$E = {-}8.31\ x\ 10{-}19\ \text{J/molecule}\ x\ 6.022.\ x\ 1023\ \text{molecules/mole}$$

The relative energy of an isolated ion is −500 kJ/mol, the standard molar enthalpy of the formation of the ion pair can be estimated by using the Coulomb law is −123 kJ/mol (-500 kJ/mole + 377 kJ/mole). Even in the case of a single Na ion and Cl ion in the gas phase, it the lower energy that is available through the association of the ions of opposite charges. This drives the formation of ionic compounds.

The ionic radii that is used in the calculation in the above reactions, were the radii of Na and Cl ion that can be found in ionic crystals. However, they are similar to the radii of these ions but under different condition. The actual calculation distance between the ion in Na+ Cl- is found to be 236.1 pm.

The association of ions that are of opposite charges are not normally into ion pairs. It is found to be easy in case of ions that are in the solid ionic crystal forms, which are consider as the three-dimensional array of the ions.

The diagram of Born-Haber cycle that is shown below, shows the formation of an ionic compound from the reaction of an alkali metals like

Li, Na, K, Rb, Cs, with the gaseous halogens like F2 and Cl2. The Born-Haber thermochemical cycle is named after two German physical chemists, Max Born and Fritz Haber. They were the first who used it in the year 1919.

## Born - Haber Cycle

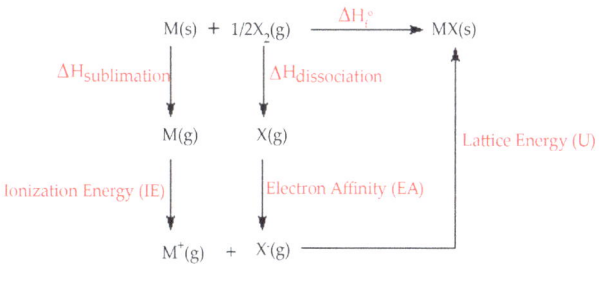

$$\Delta H_f^\circ = \Delta H_{sub} + IE + \Delta H_{diss} + EA + U$$

The enthalpy change that is in the formation of an ionic lattice, that happens from the gaseous isolation of Na and Cl ions is −788 kJ/mol. Lattice energy of the ionic crystal can be explained as the enthalpy change corresponding to the reaction Na+(g) + Cl-(g) $\longrightarrow$ NaCl(s). Although the lattice energy cannot be directly measured, there are various ways that can estimate it from the theoretical consideration and by some of the experimental values.

For the ionic crystals that are known already, their lattice energy has a large negative value. The stability and formation of ionic crystal structures, ultimately the lattice energy of an ionic crystal will be considered responsible for this.

For the sodium Cl, the Born – Haber cycle is:

$$\Delta H_f^\circ = \Delta H_{sub} + IE + \Delta H_{diss} + EA + U$$

$$\Delta H_f^\circ = 108 + 496 + 122 - 349 - 788 = -411 \text{ kJ/mole}$$

This type of cycle is a good example of Hess law. It can be used to calculate any of the six enthalpies, given the other five.

Example:

Calculate the first ionization energy IE1 for Mg atom from the following data:

ΔHsublimation Mg(s) = 130 kJ/mol

IE2 = 1441.7 kJ/mol

O2 Bond Energy = 498.7 kJ/mol

$O(g) + e- \bullet O-(g)$          ΔH = -141 kJ/mol

$O-(g) + e- \bullet O2-(g)$          ΔH = 780 kJ/mol

ΔHf MgO(s) = - 601.8 kJ

Lattice energy of MgO(s) = 3800 kJ/mol

Solution:

ΔHf MgO(s) = - 601.8 kJ

$Mg2+(g) + O2-(g) \bullet MgO(g)$

Therefore:

ΔHf O2-(g) = ½ (bond energy of O2) + IE1 of O2 + IE2 of O2

= ½ (498.7) + (-141) + 780

ΔHf O2-(g) = 888.35 kJ/mol

For Mg:

ΔHf Mg2+(g) = ΔHsublimation Mg(s) +IE1 of Mg + IE2 of Mg

= 130 + IE1 of MgO + 1441.7

Since ΔHf MgO (g) = ΔHf MgO(s) + lattice energy of MgO

Therefore:

ΔHf MgO (g) = -601.80 + 3800 = 3198.82 kJ/mol

We know that:

ΔHf MgO (g) =    ΔHf Mg2+(g) + ΔHf O2-(g)

3198.82     = [130 + IE1 of MgO + 1441.7] + 888.35

Therefore:

IE1 of MgO = 3198.82 -888.35-130-1441.7 = 738.8 kJ/mol

# 3.12. THERMODYNAMICS OF ION EXCHANGE

## 3.12.1. Ion Exchange Equilibria

At the time of ion exchange process, ion gets essentially converted from the solvent phase to solid surface state. As it is said, the process of binding ions takes place on the surface of the solid surface. Therefore, the rotational and translation freedom of the solute is said to be reduced. Therefore, the entropy change (i.e., $\Delta S$) becomes negative at the time of ion exchange. . The Gibbs free energy change ($\Delta G$) must be negative which requires the enthalpy change to be negative. This is because of the equation that takes place:

$$\Delta G = \Delta H - T\Delta S$$

Therefore, the changes in the enthalpy ($\Delta Ho$) and the entropy ($\Delta So$) helps in deciding the overall selectivity of the ion exchange process (Marcus Y., Sen Gupta A. K. 2004). Thermodynamics is said to have a great efficiency on the impulsion of the ion exchange. It also helps in setting the equilibrium distribution of the ions between the solid and the solution.

When one type of a free mobile ion of the solution becomes fixed on the solid surface, then this is included as the basic rule for ion exchange process. This process releases different kinds of ions from the solid surface. This process is irreversible process which means that there are no permanent changes that happens on the solid surface.

Environmental, medical, and technological are few of the applications of the ion exchange process that works in the different fields. To have an evaluation about the properties and efficiency of the ion exchange process, it is important to determine the equilibrium conditions.

At equilibrium conditions, there is no mass diffusion and concentration gradient through the surface of the ion exchanger. By simply plotting the concentration (CsA+) or equivalent fraction (xsA+) of A+ at the surface versus the concentration (CA+) or equivalent fraction(x+A) of this ion, the ion exchange equilibrium process can be explained.

For monovalent ions the equivalent fractions are given by:

$$x_A^{B+} = \frac{m_A^{B+}}{m_A^{B+} + x_B^{B+}} \qquad E_1$$

$$x_B^{B+} = \frac{m_B^{B+}}{m_A^{B+} + x_B^{B+}} \qquad E_2$$

In the above equation, m denotes the concentration of the specified ions at the surface. Ion exchange reactions can also be described by the law of mass action by using the activities of the ions, rather than their concentrations (Bergaya et al., 2006);

$$K \quad \frac{a_B^{B+} \quad a_A^{B+}}{a_A^{B+} \quad a_B^{B+}} \qquad E$$

In the above equation, where aA+ and aB+ are the activities of ions A+ or B+ in solution asA+ and asB+ at the surface of the exchanger. In equation E3, K represents the thermodynamic equilibrium constant. There are some common models that are used to explain the ion exchange equilibrium.

First type of this model is based on the mass action law (Gorka et al.,2008; de Lucas et al.,1992; Melis et al.,1995; Valverde et al., 1999). In the second group, ion exchange is treated as an adsorption process (Gorka et al., 2008) and in the third, the solid phase is considered to provide sites with fixed charges for ion exchange, as well as sites on which molecular adsorption takes place (Gorka et al.,2008; Novosad and Myers,1982; Myers and Byington,1986; Ioannidis and Anderko,2001).

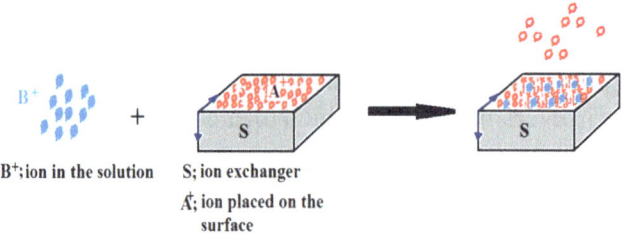

B+;ion in the solution     S; ion exchanger
A; ion placed on the surface

$$B^+ + AS \rightarrow BS + A^+ \qquad E_4$$

## 3.12.2. Ion Exchange Isotherm

The process of ion exchange equilibrium can be also assessed by the suitable equilibrium isotherms. These isotherms are the graphical representations that shows the correlation between the equilibrium and the experimental terms at constant temperature.

In solids, the concentration of an ion can be expressed as a function of its concentration. The most common ion exchange isotherm can be represented by the solidarity that takes place between the ionic composition of the two phases. These two phases are the ion exchange material and the solution (Zagorodni, 2007).

In the case of ion exchange system, selectivity is the widely used characteristic. It shows the choice of the material of one ion in comparison with another ion. Selectivity is said to be a comparative value. The easiest way of defining the selectivity is done by comparing the equivalent fractions of the ions that is in the two phases. If the equivalent fraction of this ion in the exchanger phase is higher than in the surrounding solution than the exchanger, it is considered selective towards one ion.

Such a special selection characteristic usually depends on the problem to be solved and on the experimental data available (Zagorodni, 2007). Confounding processes such as chemisorption, microbial degradation, precipitation, and steric effects, when present generates selectivity coefficients that can be used to generate exchange isotherms utilizing a reverse order of the usual procedure (Bloom and Mansell, 2001).

## 3.13. CONCLUSION

An ion is considered as an atom or molecule that contains net electric charge. This electric charge is due the loss or gain of one or more electrons from the atom. Ions can be distinguished in to two parts - namely anions and cations. Anions are formed when a greater number of electrons is gained by the atom and cations are formed when electrons leave an atom's outer shell. So, this means that anions are negatively charged and cations are positively charged particles.

The formation of ionic bonds also known electrovalent bonds are participated by the absolute transfer of electrons form one atom to another. Ionic bond is a type of linkage that is formed by the help of electrostatic attraction that takes place between the oppositely charged ions in a chemical reaction. Covalent bonds are formed when the atoms of the same or different elements combine together by mutual sharing of electrons. Ions having regular or repeating arrangements of ionic bonds produces substances with ordered arrangement or ionic lattice.

## REFERENCES

1.    Anon, (n.d.). Chemistry Handout. [online] Available at: http://www. tiwariacademy.com/wp-content/uploads/2015/10/11-chemistry-handout-chapter-4.pdf (accessed on 13 March 2020).

2.    BBC Bitesize. (n.d.). Ionic Compounds – AQA – Revision 3 – GCSE Combined Science – BBC Bitesize. [online] Available at: https://www. bbc.co.uk/bitesize/guides/ztc6w6f/revision/3 (accessed on 13 March 2020).

3.    Chemical Bonding and Molecular Structure, (2020). [ebook] Available at: http://www.ncert.nic.in/ncerts/l/kech104.pdf (accessed on 13 March 2020).

4.    Chemistry.bd.psu.edu. (n.d.). Born-Haber Formation of Ionic Compounds. [online] Available at: http://chemistry.bd.psu.edu/ jircitano/BH.html (accessed on 13 March 2020).

5.    Courses.lumenlearning.com. (n.d.). Ions | Introduction to Chemistry. [online] Available at: https://courses.lumenlearning.com/introchem/ chapter/ions/ (accessed on 13 March 2020).

6.    Encyclopedia Britannica. (n.d.). Ionic Bond | Chemistry. [online] Available at: https://www.britannica.com/science/ionic-bond (accessed on 13 March 2020).

7.    Helmenstine, A., (2019). A Covalent Bond in Chemistry is a Link between Two Atoms or Ions. [online] ThoughtCo. Available at: https:// www.thoughtco.com/definition-of-covalent-bond-604414 (accessed on 13 March 2020).

8.    Ion Exchange Chromatography Principles and Methods, (2016). [ebook] Sweden: GE Healthcare Bio-Sciences. Available at: https:// research.fhcrc.org/content/dam/stripe/hahn/methods/biochem/Ion_ Exchange_Chromatography_Handbook.pdf (accessed on 13 March 2020).

9.    Kilislioglu, A., (2012). Thermodynamics of Ion Exchange. [online] Available at: https://www.intechopen.com/books/ion-exchange-technologies/thermodynamics-of-ion-exchange (accessed on 13 March 2020).

10.   Lias, S., & Bartmess, J., (n.d.). Gas-Phase Ion Thermochemistry. [online] Webbook.nist.gov. Available at: https://webbook.nist.gov/ chemistry/ion/#IE (accessed on 13 March 2020).

11.   Marshall, K., (2017). What is Ion Exchange Resin and How Does it

Work? [online] Samco Tech. Available at: https://www.samcotech. com/ion-exchange-resin-work-process/ (accessed on 13 March 2020).

12. Mendoza, RMO, Kan, C-C, Dalida, MLP, Wan, M-W (2018). Groundwater treatment by electrodialysis: Gearing up towards green technology. Desalination and Water Treatment, 127, pp. 178-183.

13. Mendoza, RMO, Kan C-C., Chuang, S-S, Pingul-Ong,SM., Dalida, MLP, Wan, M-W (2014). Feasibility of arsenic removal from aqueous solutions be electrodialysis. J.Envi.Sci and Health, Part A Toxic/ Hazardous Substances and Environmental Engineering

14. The Cavalcade o' Chemistry, (2015). Lewis Structures of Ions. [online] Available at: https://chemfiesta.org/2015/01/26/lewis-structures-of-ions/ (accessed on 13 March 2020).

15. Walton, H., (n.d.). Ion-exchange Reaction | Chemical Reaction. [online] Encyclopedia Britannica. Available at: https://www.britannica.com/ science/ion-exchange-reaction (accessed on 13 March 2020).

# Kinetic Molecular Theory of Gases

## CONTENTS

This chapter discusses in detail the Kinetic Molecular Theory (KMT) of gases. The discussion focuses mainly on the molecular and atomic explanations of how gases behave ideally and in the actual experimental set-up. The discussions start with san overview of KMT, followed by related gas laws. The application of KMT on different experimental conditions as well as in the dynamics of gas molecules in each related unit process is also tackled.

## 4.1. INTRODUCTION

Kinetic theory of gases, a theory based on a simplistic definition of a gas by molecules or particles from that many gross properties of the gas can be extracted. The British scientist James Clerk Maxwell and the 19th century Austrian physicist Ludwig Boltzmann led the establishment of the theory, that has become one of the most key concepts in modern science particularly in physics and chemistry.

The simplest kinetic model is based on the premises that:

1.     The gas is composed of a large number of similar molecules moving in random directions, segregated by large distances relative to their size;

2.     The molecules are subjected to perfectly elastic collisions-that is no loss of energy with each other and with the container walls, but they do not interact otherwise; and

3.     The transfer of kinetic energy (KE) between molecules is heat. Such simplifying hypotheses put gas characteristics within the range of mathematical treatment.

This type of model defines a perfect gas and is a good approximation to an actual gas, particularly in the extreme dilution limit and/or high temperature. Nevertheless, such a simplistic definition is not sufficiently accurate to account for the behaviour of high-density gases.

On the basis of theory of kinetics, pressure on the container walls can be quantitatively related to spontaneous collisions of molecules of which the average energy relies on the gas temperature. Therefore, the gas pressure could be directly related to temperature and density.

The gas can derive many other properties, such as viscosity, thermal, and electrical conduction, diffusion, heat capacity, and mobility. The premises must be properly adjusted to clarify observed deviations from ideal gas behaviour, such as condensation and another phenomenon. In doing so, the nature of molecular dynamics and interactions has gained considerable insight.

Gases can be analysed by considering the small-scale behaviour of individual molecules or by considering the overall large-scale action of the gas. The gas' large-scale behaviour can be directly measured, or estimated. Though one should use a theoretical model to analyse the behaviour of the molecules.

The model, called the kinetic theory of gases, states that due to the distance between molecules, the molecules are extremely small. The molecules are in continuous, erratic motion and often collide with each other and within the  walls of the container.

The standard physical properties of mass, momentum, and energy are retained by the individual molecules.

A density of a gas is essentially the sum of the molecular mass divided by the volume that the gas occupies. The pressure from a gas is a function of the molecule's linear momentum.

The molecules add energy to the walls as the gas molecules collide with the walls of a container, creating a force The  temperature of the gas is a function of the gas' mean KE. The molecules are in continuous random motion, and that motion is correlated with an energy (mass x square of the velocity). The higher the temperature, the greater the motion.

The position of the molecules in a solid remains nearly constant relative to each other. But in a gas, the molecules will travel around and interact in different ways with each other and their surroundings. As described, molecular motion always has a random component. It can make the entire fluid travel in an orderly motion (flow) as well.

The organized motion is superimposed over or applied to the molecules' normal random motion. There is no distinction, at the molecular level, between the random component and the ordered component. The static pressure calculates the pressure produced by the random variable.

The pressure arising from the ordered motion is called dynamic pressure. And Bernoulli's equation tells us that the total pressure which we can also calculate is the sum of the static and dynamic pressure.

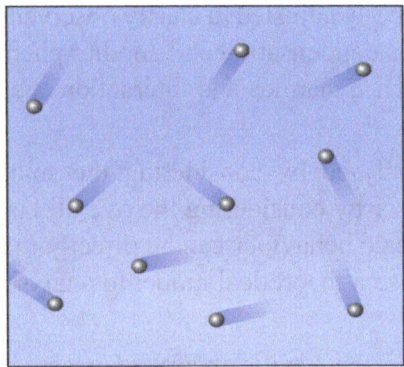

**Figure 4.1**. Kinetic theory of gases.

*Sources: Image by Wikipedia.*

## 4.2. EARLY THEORIES OF GASES

The kinetic theory of gases emerged from the ancient belief that matter consists of small, fast moving invisible atoms. The idea was reignited in the 17th century, and used to describe the properties of gases, among other phenomena. Building on the work of several other 17th-century scientists, the British chemist and physicist Robert Boyle (1627–1691) proved that air is elastic: it resists compression and spreads to fill the space available.

The mechanical pressure P applied by a given quantity of gas at a specific temperature is inversely proportional to the volume V of its container, a relationship now known as Boyle's Law. Boyle stated two alternative atomistic theories for air pressure:

1.   Air consists of particles that repel each other, such as coiled-up bits of wool or springs; and

2.   Air consists of whirling particles that push each other away by impacts. Isaac Newton took up the first assumption, that proved mathematically that if air pressure is generated by the repulsion of adjacent particles, then the repulsive force must be inversely proportional to their distances.

In the 17th century, Boyle's second theory identified with the etheric vortices of Descartes lacked a quantitative basis, although it received qualitative support from the common idea that heat is correlated with atomic motion and the observation that air pressure rises with temperature.

Within his book on hydrodynamics, the Swiss mathematical physicist Daniel Bernoulli (1700–1782) proposed a systematic theory of kinetics. He derived Boyle's gas pressure law by measuring the force exerted on a moving piston by the effect of *n* particles moving at a certain velocity, in a closed space of total volume, V.

When V is lower, the pressure will be higher because the particles hit the piston more often. If the space occupied by the particles themselves is small compared to the volume V, the pressure P should be inversely proportional to V; thus, as Boyle's law states, the product PV is constant (Bernoulli, 1738).

Bernoulli also demonstrated that the pressure will be proportional to the particle KE ($1/2mv2$ in which m is the mass of a single particle) as the impact frequency is proportional to the velocity and the force of each impact is proportional to the mv momentum.

This, as he concluded, he explained the observed fact that pressure increases resulting from equal temperature increases are proportional to the density, and proposed that temperature could be defined at a standard density in terms of air pressure itself.

Although the definition of an absolute temperature scale had not yet been embraced by other scientists, Bernoulli's theory presented the idea that heat or temperature can be measured in an ideal gas with the KE of particles. Experimental work on gases around 1800 established the Bernoulli assumed simple relationship between pressure, volume, and temperature.

The French chemist Joseph Gay Lussac (1778–1850) and others have shown that pressure increases in proportion to temperature if the volume is kept constant, or volume increases in proportion to temperature if the pressure is kept constant; these relations can be summarized in equation $PV = NR(T+273)$, where N is proportional to the total mass of gas present, T is temperature in degrees Celsius.

But it was not yet known whether the equation will be valid down to such low temperatures that (T + 273) would be zero, or whether all gasses would condense before that absolute cold point would be reached so that the equation would no longer apply. In the 18th century the kinetic principle was not widely accepted; most scientists favoured the Newtonian theory of repulsion, that was consistent with the belief that heat is a fluid, caloric, instead of the energy of atomic motion.

Often, caloric was assumed to be composed of particles repelling each other and drawn to the atoms of ordinary matter. Therefore, gas pressure increases with temperature because more of the self-repelling caloric fluid is acquired by the gas. Temperature could itself be defined as caloric density (the sum of caloric fluid divided by volume).

With that kind of definition of temperature, the caloric theory could explain why compression can increase a gas's temperature even though no outside heat is added (the same quantity of caloric is concentrated in a smaller volume), or expansion can decrease the temperature even if no heat is lost (the same quantity of caloric is spread over a greater volume). While there was one anomalous discovery whose purpose was not understood until much later: Gay-Lussac discovered that there is virtually no change in temperature in the free expansion of a gas into a vacuum rather than pushing back a piston.

The caloric theory can also explain occurrences like the latent heat of phase transitions (solid to liquid or liquid to gas) and the heat absorbed or released in chemical reactions, by pretending that some caloric is bound to the individual atoms or compounds. The ordinary volume-pressure relationship of gases is decided by the unbound or free caloric which fills the space between particles.

The kinetic theory seemed to give no logical account of these phenomena, and moreover its assumption that the atoms pass between collisions at constant speed seemed inconsistent with the generally accepted belief that all space is filled with an etheric fluid.

Finally, in the early 19th century, caloric theory gained legitimacy from Laplace's use of it to measure the velocity of sound in gases, overcoming a long-standing disparity between theory and observation; and it was partially endorsed by the acceptance of the particle theory of light, because light and heat were widely regarded as qualitatively equivalent phenomena.

## 4.3. MOLECULAR NATURE OF MATTER

One of the great scientists of the 20th century, Richard Feynman, agreed by the observation that Matter is made-up of atoms to be a very important.

If humankind don't behave wisely, civilization will experience annihilation because of nuclear catastrophe or extinction due to environmental catastrophes. If that occurs and all of the scientific knowledge is to be lost then Feynman would like to transmit the Atomic Hypothesis to the next generation of beings in the world.

## 4.3.1. Atomic Hypothesis

All things are made up of atoms-small particles traveling in perpetual motion, attracting each other when they are a little distance apart, but repelling each other when they are crammed into one another. There may not be continuous theories that matter but existed in many places and cultures. In India - Kanada, and in Greece - Democritus had proposed that matter could consist of indivisible constituents (though John Dalton was s credited with the scientific atomic theory).

Dalton presented the theory of atomics to describe the rules of definite and multiple proportions that elements obey as they combine to form compounds. The first law says that there is a fixed proportion of any compound by mass of its constituents. The second law states that when two elements form more than one compound, the masses of the other elements are in the ratio of small integer for a fixed mass of one element.

In order to demonstrate the laws, Dalton proposed that the smallest constituents of an element are atoms, about 200 years ago. Atoms of one element are the same but are different from those of other elements. A small number of atoms of each element join together to form a compound molecule.

Gay Lussac's law, also laid in the early 19th century, states that: When gasses chemically combine to produce another gas, their volumes are in the small integer ratios.

Avogadro's then hypothesized that The Equal quantities of all gasses have the same number of molecules at equal temperature and pressure. Now called Avogadro's law, it was combined with Dalton's theory of which explains of Gay Lussac's law. As the elements are mostly in the form of molecules, it is also possible to refer to Dalton's atomic theory as the molecular theory of matter. Scientists are now embracing the idea in a good way.

Even at the end of the 19th century, however, there were prominent scientists that did not believe in the theory of atoms. In recent times, scientists now learn from many experiments that molecules consisting of one or more atoms constitute matter. Electron microscopes and tunnel scanning microscopes also help us to see them.

One atom's size is about an angstrom (10 −10 m). Atoms are spaced only a few angstroms (2 Å) apart in solids, which are densely packed. The difference between atoms is about the same also in liquids. The atoms are not as rigidly set in liquids as they are in solids and can move around.

This allows liquids to flow. The interatomic distances in gases are within tens of angstroms. The mean free path is called the average distance that a molecule can travel without colliding. For gases the mean free path is of the order of thousands of angstroms.

The atoms in gases are much freer, and they can travel long distances without colliding. Gasses scatter away if they are not sealed. The closeness of the interatomic force makes it essential in solids and liquids. The force has an attraction for the long range and a repulsion for the short range.

When they are at a few angstroms, the atoms attract but repel when they get closer. A gas is misleading in its static appearance. The gas is full of movement, and complex equilibrium. During the collision molecules collide in dynamic equilibrium and change their velocities.

Only the average attributes are unchanged. Atomic theory is not the end but the beginning of our search. They now know that neither indivisible nor elementary atoms are. They are created by a nucleus and electrons. The nucleus itself is composed of neutrons and protons.

The quarks again comprise the protons and neutrons. But quarks are perhaps not the end of the story. The string could be like elementary entities. Nature has always surprised for us, but the quest for truth is often fun, and the discoveries are beautiful.

## 4.4. BASIC CONCEPT OF GAS

Gases were among the first substances that were researched using the modern scientific method, at was established around 1600. It didn't take long to realize that all gasses shared similar physical characteristics, implying that one all-encompassing principle could explain gases. The kinetic molecular theory (KMT) of gases is a model that allows us to understand at the molecular level the physical properties of gases.

It takes the following concepts as its basis:

1.  Gases are composed of particles molecules or atoms that are in continuous random motion.
2.  Gas particles clash continuously with one another and with the wall of container. These collisions are elastic; that is, the collisions do not cause a net loss of energy.
3.  Gas particles are small, and the total volume taken up by gas molecules is miniscule in relation to their total container volume.

4. There are no interactive forces, i.e., attraction or repulsion exist between a gas's particles.

5. The average KE of gas particles is proportional to the absolute gas temperature, and all gases have the same average KE at the same temperature.

The kinetic molecular gas theory defines this state of matter as being composed of small, constantly moving particles with a lot of space between the particles. Because most of the volume a gas occupies is empty space, a gas has a low density and under the appropriate influence can expand or contract. The fact that gas particles are in constant motion means that two or more gasses often interact with each other as the particles from the individual gasses travel and collide.

The number of collisions that the gas particles make with their container walls, and the force they collide with, decide the scale of the gas pressure.

The history of the nature of gas is a very vast field. From experiments that are too numerous to go into here and from the theory that s presented scientist the human beings know that an ordinary volume of gas consists of billions of molecules. At normal temperatures the molecules travel at high velocities in all directions.

They deduce that if it could be observed, the motion of any one molecule would be quite random. The molecules do not all travel at the same speed at any time, but with a spectrum or distribution of speeds, certain speeds are somewhat more sluggish.

## 4.4.1. Some Important Properties of Gases

Apart from liquids, any gas still mixes in any proportion, i.e., they are miscible and form homogeneous solutions thoroughly with any other gas.

• Gasses Are Compressible

The volume of the gas decreases when the pressure is applied. It is relatively incompressible for liquids and solids.

• The Volume of a Gas Expands with Heating and Contracts on Cooling

For gases, that effect is much greater than for liquids or solids.

• Gasses Have Relatively Low Viscosity

It means that they float much freer than liquids or solids. Under normal conditions, most gases are of relatively low densities. (Oxygen: 1.3 g/L, 2.2 g/mL NaCl)

## 4.4.2. Gas Pressure and Its Measurement

A gas consists of particles in a medium that is essentially empty space, moving at random. Pressure is the force that gas molecules exert per unit of area, while they hit the surfaces around them. Not every single collision exerts a lot of force, but the total force exerted by the large number of particles in the gas adds up to a great force.Gas pressure is measured using a barometer that is consisting of a long tube that is sealed at one end and filled with mercury and inverted into a Hg platter. Some Hg flows out of the tube until the atmospheric pressure on the Hg in the platter balances the downward pressure of the Hg in the dish.

At sea level and 0°C the Hg column is 760 mm in height. In Denver (altitude ~1 mile), a column of Hg is 630 mm tall–a column of Hg is 270 mm tall at the top of Mt. Everest (29,028 ft.)The instrument used for measuring pressure is a manometer. Any instrument that calculates pressure may be a manometer. Unless otherwise eligible, however, the term manometer most often refers specifically to a tube in U-shaped form partially filled with fluid. In order to demonstrate the effect of air pressure on a liquid base, this type of manometer can be easily built as part of a laboratory experiment.

### 1.    Building a Manometer

A simple manometer can be built by partially filling a clear plastic tube with a coloured liquid to allow for easy measurement of the fluid level. Afterwards, the tube is bent into a U-shape and set upright. At this point the fluid levels in the two vertical columns should be equal, as they are actually being subjected to the same pressure. Therefore, that amount is labelled and defined as the manometer's zero point.

### 2.    Pressure Calculation

The manometer is set against a calibrated scale to permit any difference in the height of the two columns. This difference in height can be used explicitly to make comparisons between different test pressures. Often, this type of manometer can be used to measure the absolute pressure when the liquid density is established in the manometer.

### 3.    The Way It Works

One end of the tube is attached to a test pressure source using a gas-tightened seal. The other end of the tube is left open to the atmosphere and is thus subjected to a pressure of about 1 atmosphere (atm). If the test pressure reaches the reference pressure of 1 atm, then the liquid in the test column will be forced down the tube. That causes the fluid to rise by an equal amount in the reference column.

#### 4.    Pressure Measurement

The pressure exerted by a fluid column can be measured by equation P = hgd. In that equation, P is the calculated pressure, h is the fluid height, g is the gravity force, and d is the liquid density. Because the manometer calculates a differential pressure rather than an absolute pressure, they use the alternative P = Pa – P0. Pa is the check pressure in this replacement, and P0 is the reference button.

**Figure 4.2.** A manometer is a scientific instrument used to measure gas pressures.

*Sources: Image by Wikipedia.*

### 4.4.3. General Gas Law

The behaviour of a gas under different temperature and pressure conditions has already been studied. When the pressure of a constant mass of gas isn't too high, say less than about 2 atm, they consider a gas obeys the following relationships: at constant temperature PV= constant; at constant volume P= KT; at constant pressure V= KT.

All three equations are special cases of a single experimental equation that gives the relationship of a constant mass of gas between the pressure P, volume V, and absolute temperature T.

### 4.4.4. The Gas Laws Explained from a KMT Perspective

- Kinetic Interpretation of Boyle's Law: Boyle's law is easily explained through the theory of kinetic molecules. The strength of a gas depends on the number of times a second the molecules target the surface of container. When one compresses the gas to a smaller volume, the same number of molecules now work against a smaller surface area, so that the number striking per unit of area, and hence the strain, is greater now.

- Kinetic Interpretation of the Law of Charles: KMT states that the temperature increase increases the average KE of the molecules. If the molecules move faster but the pressure remains the same, then the molecules have to stay further apart so that the change in the rate at which the molecules collide with the container surface is compensated for by a corresponding increase in the area of this surface as the gas expands.

- Avogadro's Kinetic Explanation: When they increase the number of gas molecules in a closed container, more of them collide with the walls per unit time. The volume must increase in proportion if the pressure is to remain constant, so that the molecules strike the walls less frequently, and over a larger surface area.

- Dalton's Law's Kinetic Explanation: Every gas is a vacuum for every other gas. This is the way Dalton announced what they now consider to be his rule of partial pressure. This simply means that every gas that is present in a gas mixture acts independently of the others. This makes sense because the first fundamental principle of KMT theory is that there are negligible volumes in the gas molecules. But Gas A in mixture of A and B behaves as if there was no Gas B at all there. Each contributes its own pressure to the total container pressure, proportional to the fraction of the molecules it represents

## 4.5. THE FUNDAMENTALS OF KINETIC MOLECULAR THEORY (KMT)

A gas's molecules are in a state of perpetual motion in which each molecule's velocity that is, velocity, and direction is entirely random and independent of that of the others.

The fundamental assumption of the kinetic-molecular model helps us understand a wide variety of phenomena commonly observed. According to this model, most of the volume filled by a gas is empty space; this is the main characteristic that distinguishes gases from condensed state of matter liquids and solids in which adjacent molecules are constantly in contact.

Gas molecules are in rapid and continuous motion; their velocities are of the order of 0.1–1 km per sec at ordinary temperatures and pressures and each molecule experiences about 1010 collisions with other molecules per second.

The Five Fundamental Tenets of Kinetic-Molecular Theory Are:

1.    A gas consists of molecules that are segregated by average distances much larger than the sizes of the molecules themselves. The volume that the gas molecules occupy is insignificant compared with the gas volume itself.

2.    The ideal gas molecules do not exert any attractive forces on each other, or on the walls of a container.

3.    The molecules are in continuous random motion, and they obey Newton's laws of motion as material bodies. That means the molecules move in straight line until they collide with one another or the container walls.

4.    Collisions are completely elastic; they change their paths and kinetic energies as two molecules collide, but the overall KE is conserved. Collisions are not sticky.

5.    `The average KE of the gas molecules is directly proportional to the absolute temperature.

Here the word average is very important; the velocities and kinetic energies of the individual molecules will span a broad range of values, and some will even have zero velocity at a given moment. This means that all molecular motion would cease if the temperature were decreased to absolute zero.

## 4.5.1. Assumptions of the Kinetic Molecular Theory

The kinetic theory involves a number of assumptions that focus on being able to talk about an ideal gas.

•    Molecules are viewed as point particles

•    One consequence of this is precisely that their size is extremely small compared with the average particle diameter.

•    The number of molecules (N) is very high, in so far as it is not possible to track individual particle behavior. Statistical methods are used to evaluate the system's actions as a whole, instead.

•    Each molecule shall be treated as the same as any other molecule.

As to their different properties, they are interchangeable. This again helps to support the concept that there is no need to keep track of individual particles and that the theory's statistical methods are enough to arrive at conclusions and predictions.

- Molecules are in constant, arbitrary motion. They follow Newton's motion laws.
- Collisions are fully elastic collisions between the particles and the walls of the container for a gas.
- Gas container walls are viewed as completely solid, do not move, and are infinitely massive as contrasted with particles.

The result of these assumptions is that inside a container you have a gas that moves around within the container randomly. If gas particles collide with the side of the container, they bounce off the side of the container in a perfectly elastic collision, ensuring they bounce off at an angle of 30 degrees if they strike.

Their velocity part perpendicular to the side of the container changes direction but maintains the same magnitude.

## 4.5.2. Kinetic Theory of an Ideal Gas

The kinetic-molecular theory accounts for the atoms and molecules behavior based on the idea that all matter particles are always in motion. This theory gives us a model of what is known as an ideal gas, an imaginary gas that matches all KMT assumptions perfectly.

The theory of kinetic gases is based on the molecular image of matter. A given quantity of gas is a set of a large number of molecules generally in the order of the number of Avogadro that are in continuous random movement. The mean distance between molecules at ordinary pressure and temperature is a factor of 10 or more than the typical size of a molecule (2 Å).

Hence the interaction between the molecules is minimal and according to Newton's first law we can assume that they move freely in straight lines. Occasionally, however, they get close to each other, feel intermolecular forces and adjust their velocities. Such interactions are denominated collisions.

Incessantly the molecules collide with each other or with the walls and change their speeds. The collisions are deemed to be elastic. Based on kinetic theory, they can derive an expression for the pressure of a gas. They continue with the notion that a gas's molecules are in continuous random motion, colliding with each other and the walls of container.

Both molecular collisions between themselves or between the molecules and the walls are elastic. This means retaining of complete KE. The overall energy is normally maintained.

## 4.5.3. Pressure of an Ideal Gas

Consider a gas enclosed in a cube of side l. Take the axes to be parallel to the sides of the cube, as shown in Figure 4.3. A molecule with velocity

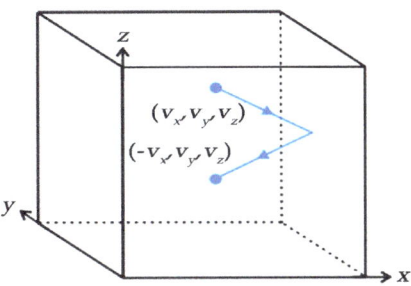

**Figure 4.3**. 3-dimentional view of gas motion

(vx, vy, vz) hits the planar wall parallel to yz-plane of area A (= l2).

Because the collision is elastic, the molecule rebounds at the same speed; its velocity components y and z do not change in the collision, but the x-component reverses the sign. That is, after collision the velocity is (-vx, vy, vz). The change in momentum of the molecule is: –mvx–(mvx)= –2mvx.

By the theory of conservation of momentum, the momentum given to the collision wall= 2mvx. To measure the force and pressure on the wall, the momentum imparted to the wall must be measured per unit time. A molecule with x-component of velocity vx will hit the wall if it is within the vxΔT distance from the wall within a small-time interval ΔT.

That is, in time, all molecules within the A vx ΔT volume can only hit the wall. Yet, on average, half of these travel towards the wall, and the other half away from the wall, Accordingly, the number of molecules with velocity (vx, vy, vz) hitting the wall in time ΔT is ¬A vxΔT n where n is the number of molecules per unit volume.

The total momentum transmitted by these molecules to the wall in time T is: Q= (2mvx) (¬ n A vx T) The force on the wall is the rate of momentum transfer QΔT and the pressure is force per unit area: P= Q/(AΔT)= n m vx2 In fact, all molecules in a gas do not have the same velocity; there is a variation in velocity. Therefore, the above equation stands for pressure because of the group of molecules with speed vx in the x-direction and n represents the number density of that group of molecules.

## 4.6. SOME IMPORTANT PRACTICAL APPLICATIONS OF KMT: THE ROOT MEAN SQUARE (RMS)

A molecule of the gases in a state of constant motion in which each molecule's velocity that is velocity and direction is entirely random and independent of that of the other molecules. The basic assumption of the kinetic-molecular model helps to learn about a wide range of phenomenon commonly observed.

Since gas molecules are generally in random motion, this means that they are moving in all directions. Thus, each molecule with random speed in one direction is cancelled out by another molecule randomly moving in the exact opposite direction (with the same magnitude of speed). Therefore, the average velocity of gas molecule in a sample is theoretically zero.

The Root Mean Square (RMS) measures the speed of the gas molecules in relation to its average kinetic energy. As discussed in the previous section, the random motion of the gas molecules is directly proportional to the average kinetic energy. The $vRMS$ is a familiar term we see in the average kinetic energy equation, $KE = \frac{1}{2}$ m$(vRMS)$2, since the $vRMS$ is the average velocity of gas when average KE is considered. Substituting the formula for:

$$v_{RMS} = \sqrt{\frac{3kT}{m}}$$

(4.1)

in the average KE equation:

$$KE = \frac{1}{2}m(\sqrt{\frac{3kT}{m}})^2 = \frac{1}{2}m(\frac{3kT}{m})$$

(4.2)

$$KE = \frac{3}{2}kT$$

(4.3)

Equation 4.3, therefore, indicate that the average kinetic energy of gases is not affected my its mass. On the other hand, the relationship between $k$ and R is that $k = R/M$ thus, equation 4.1 can also be written as:

$$v_{RMS} = \sqrt{\frac{3RT}{M}}$$
                                                                    (4.1.1)

For the ease of using equation 4.1.1, the unit of M must be converted to kg/mols of gas and R=8.314 J/mol K.

Example:

1.  Calculate the speed of oxygen gas molecule at 38°C.

Given:

T = 38+273.15 = 311.15 K

M = 32 g/mol x (1 kg/1000 g) = 0.032 kg/mol

Required: $v_{RMS}$

Solution:

$$v_{RMS} = \sqrt{\frac{3(8.314)(311.15)}{(0.032)}} = 492.47 \text{ m/s}$$

(In chemistry, the distinction between speed and velocity if often times ignored specially in terms of gas molecules)

2.    At room temperature, which will reach your lungs faster, carbon monoxide or oxygen? At what rate?

Given:

T = 25+273 = 298.15 K

$MO2$ = 32 g/mol = 0.032 kg/mol

$MCO$ = 28 g/mol = 0.028 kg/mol

Solution:

$$v_{RMS(O_2)} = \sqrt{\frac{3(8.314)(298.15)}{(0.032)}} = 482.07 \text{ m/s}$$

$$v_{RMS(CO)} = \sqrt{\frac{3(8.314)(298.15)}{(0.028)}} = 515.35 \text{ m/s}$$

*Carbon Monoxide will travel faster towards the lungs. This is why they say, carbon monoxide is a "silent killer" because it travels faster than oxygen gas. It competes with oxygen gas towards the lungs.

$$\frac{v_{RMS(CO)}}{v_{RMS(O_2)}} = \frac{515.35}{482.07} = 1.07$$

## *Carbon dioxide is 1.07 times faster than oxygen. 4.6.1. Diffusion: Random Motion with Direction*

Diffusion is described as the movement of matter through a gradient of concentration; the principle is that substances pass or tend to move from higher concentration regions to lower concentration ones. The spreading of tea from a tea bag into H2O, or of a person's perfume inside the room, are common examples of diffusion; one would not expect to view either process to happen in reverse.

At first it might seem strange that the arbitrary motions of molecules in their ultimate distribution can lead to a completely predictable drift. The approach to this obvious paradox is to differentiate between a person and the population. Although one cannot tell anything about an individual molecule's fate, the action of a large collection of population of molecules is subject to rules of statistics.

This is exactly analogous to the way in which actuarial insurance tables can accurately predict people's average lifespan at a given age but do not provide any detail about the fate of any individual person.

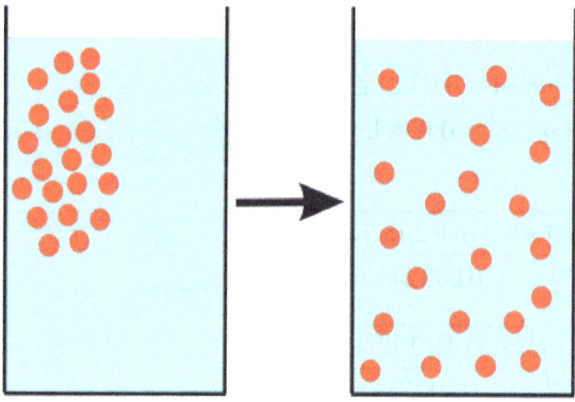

**Figure 4.4.** Process of diffusion.

*Sources: Image by Wikimedia.*

## 4.6.2. Uranium Enrichment

One implementation of this theory that Graham himself originally suggested but was not actually known till a century later is the isolation of isotopes. The uranium enrichment in nuclear fission fuel production is the most important example of this. Throughout the second World War the K-25 Gaseous Diffusion Plant was one of the major sources of enriched uranium.

It was completed in 1945 and recruited 12,000 employees. The women that worked on the device were unaware of the plant's intent because of the secrecy of the Manhattan Project; they were trained to simply watch the gauges and turn the dials for what they were told was a government project. Uranium mainly consists of U238, with only 0.7% of the fissionable U235 isotope.

Typically, uranium is a metal, but it reacts with fluorine to form UF6, a gaseous hexafluoride. The UF6 diffuses repeatedly through a porous wall in the extremely successful cycle of gaseous diffusion process. The lighter isotope moves a little faster each time than the heavier one, producing a mixture that in U235 is minutely richer.

To achieve the desired degree of enrichment, the cycle must be over a thousand times. A large-scale diffusion plant construction was a crucial part of the production of the first atomic bomb in 1945 by united states of America. This process is now outdated, with other methods being replaced.

**Figure 4.5.** Women working on Manhattan project.

*Sources: Image by Flickr.*

## 4.6.3. Evaporation

In this section, some more applications of kinetic theory have been discussed. The main feature of what is discussed below is the fact that the probability of finding a particle in different places, per unit volume, varies as e–potential energy/kT; scientists have made a number of applications of this.

The processes that scientist want to study are relatively complicated: a liquid evaporation, or electrons flowing out of the surface in a metal, or a chemical reaction that involves a great number of atoms. In such situations, clearly, and accurate assumptions can no longer be made from the kinetic principle, because the situation is too complex.

The concept to be stressed is only that they can more or less understand the way things ought to behave from the kinetic theory. One can achieve a more accurate representation of the phenomenon by using thermodynamic arguments, or some analytical measurements of certain essential quantities.

Nevertheless, it is very useful to know even more or less why something behaves as it does, so then one can assume, more or less, what should happen when the case is a new one, or one that they have not yet begun to examine. So, this discussion is highly inaccurate but essentially correct right in idea but, let's say, a little simplified in the specific details.

The first example that shall one thought is the evaporation of a liquid. Suppose one have a large volume vessel, partially filled with liquid in equilibrium, and at a certain temperature with the vapor. There is a presumption that the molecules of a vapor are relatively far apart, and the molecules are packed close together inside the liquid.

The problem is in finding out the number of molecules there are in the form of vapor, compared to the number in the liquid. The level of denseness of the vapor at a given temperature and the way does the temperature depend on that.

Let consider that n equals the number of molecules in the vapor per unit volume. Obviously, the number varies with the temperature. The more evaporation one will get if he adds heat. Now let another quantity, $1/V_a$, be equal to the number of atoms per unit volume in the liquid: One assume that each molecule in the liquid occupies a certain volume, so that if there are more molecules of liquid, they occupy a larger volume together.

Therefore, if $V_a$ is the volume occupied by one molecule, a unit volume divided by the volume of each molecule is the number of molecules in a unit volume. In fact, they assume there is an attraction force between the molecules to bind them together in the liquid.

They cannot grasp why this condenses otherwise. So, assume there is such a force and there is a binding energy of the molecules in the liquid that is lost when they go into the vapor. That is, that one is going to assume that a certain amount of work W has to be done to take a single molecule out of the liquid into the vapor.

In the energy of a molecule in the air, there is a certain difference, W, from what it would have if it were in the gas, because they have to pull it away from the other molecules that absorb it. That's why one can use the general principle that

$n2/n1 = e-(E2-E1)/kT$

is the number of atoms per unit volume in two distinct regions.

Thus, the number n per unit volume in the vapor is equal to

$nVa = e-W/kT$,

divided by the number

$1/Va$ per unit volume in the liquid, because that is the general rule.

It is like the atmosphere under gravity in equilibrium, where the gas at the bottom is denser than at the top due to the work mgh needed to lift the gas molecules up to the height h. The molecules in the liquid are denser than in the vapor, because one has to pull them out through the hill W energy, and the density ratio is $e-W/kT$.

This is what they needed to infer that the density of vapours ranges from e to the minus some energy over kT or other. The considerations before us are not particularly important, because in most situations the density of the vapor is much lower than the density of the liquids.

In those cases, where they are not close to the critical point where they are almost the same but where the density of the vapor is much smaller than the density of the liquid, then the fact that n is much lower than $1/Va$ is induced by the fact that W is much greater than kT.

So formulas like are only interesting when W is much larger than kT, because in those circumstances, as they are raising e to minus a tremendous amount, if one change T a little bit, that tremendous power changes a little bit, and the change produced in the exponential factor is much more important than any change that might occur in the factors in front of us.

The studies by scientist are rough on this subject till now. After all, for each molecule there is not really a definite amount; when one changes the temperature, the volume Va does not stay constant the liquid expands.

There are other minor things like that, so it's more complex than the actual situation. The temperature-dependent factors differ gradually all over the region.

In addition, one might conclude that W itself varies slightly with temperature, because there would be different average attractions at a higher temperature, at a different molecular volume, and so forth. So, while one may thought that if they have a formula in which everything varies unknowingly with temperature then one have no formula at all, if he realize that the exponent W/kT is generally very large, we see that in the vapor density curve, as a function of temperature, the majority of the variation is caused by the exponential factor, and if we take W As a constant and the coefficient 1/Va as nearly constant, it is a good approximation for short intervals along the curve. In other words, the bulk of the variability is of the general nature e$-$W/kT.

It has been revealed there are many, many anomalies in nature that are characterized by having to borrow an energy from somewhere, and where the central feature of the change in temperature is e to minus the energy over kT. This is a useful finding only when the energy is large in comparison to kT, so that the bulk of the variance that is found in the kT variation and not in the constant and in other variables.

The other way to obtain a somewhat similar evaporation result has been achieved in a study carried out and now it is being stated here. To come to this conclusion the rule is simply applied that is true at equilibrium, but there is no harm in trying to look at the details of that is going on in order to better understand things.

They may also explain what is happening in the following manner: the molecules that are in the vapor continuously attack the surface of liquid; they may bounce off when they touch it, or they may get trapped. There is an unknown factor one does not know for that it might be 50–50, maybe 10 to 90. Let's assume they always get stuck on the premise that they do not always get stuck that will be examined later.

Then there will be a certain number of atoms condensing onto the liquid surface at a given moment. The number of condensing molecules, the number that arrives on a unit area per unit time, is the number n per unit volume times the velocity v. This velocity of the molecules is related to temperature, since they know that 12mv2 is on average equal to 32kT.

So, v is some sort of a mean pace. Of course, they can integrate over the angles and get some sort of an average, but within some factor it is about proportional to the root-mean-square velocity. So, $N_c = nv$ is the rate at which the molecules arrive and are condensing per unit area.

However, at the same time the atoms in the liquid are wiggling around and one of them gets kicked out from time to time. Instead, one need to guess how easily they get kicked out. The concept will be that the number that are kicked out per second and the number that arrives per second will be equal at equilibrium.

The amount that get thrown out. In order to get kicked out, a single molecule has to gain an excess energy over its neighbours by accident, a significant excess energy, because it is very highly attracted by the other molecules in the liquid. It does not usually leave because it is so deeply attracted, but sometimes one of them gets an extra energy by mistake in the collisions.

And if $W \gg kT$, the likelihood it gets the extra energy $W$ it needs is very slim in this situation. In fact, $e^{-W/kT}$ is the chance an atom has gathered more energy than this much energy. That is the general principle in kinetic theory: the chances are e to the minus the energy they have to borrow, over $kT$, in order to borrow an excess energy $W$ over the average.

Now presume that some molecules have borrowed this energy. Now, they need to estimate the number of people leaving the surface every second. Moreover, only because a molecule has the requisite energy does not mean that it will necessarily evaporate, because it can be buried too deep within the liquid or, even if it is close to the surface, it can move in the wrong direction.

The number that will leave a unit area per second will be something like this: the number of atoms close to the surface, per unit area, divided by the time it takes one to escape, calculated by the probability of $e^{-W/kT}$ being able to escape in the sense of having enough energy.

They shall assume that each molecule at the liquid surface occupies a certain cross-sectional region A. Then the number of molecules shall be $1/A$ per unit area of the liquid surface. And now, how long does a molecule take to escape? If the molecules seem to have the same average speed v, and have to pass, say, one molecular diameter D, the first layer thickness, then the time it takes to get over that thickness is the time it takes to escape if the molecule has enough energy. The time is set to be $D/v$. Thus, the amount evaporating should be about $N_e = (1/A)(v/D) e^{-W/kT}$. At present the area of each atom times the thickness of the layer about the same as the volume $V_a$ occupied by a single atom. In a way to get balance they must have $N_c = N_e$, or $nv = (v/V_a) e^{-W/kT}$.

They might cancel the v's as they are equal; although one is the velocity of molecule in the vapor and the other is an evaporating molecule's velocity, they are the same because they know their mean KE in one direction is 12kT.

These are the ones that are particularly fast-moving; these are the ones that have gathered excess energy. It is not in reality because the moment they start pulling away from the liquid, they have to lose that excess energy against the potential energy. And when they get to the surface, they get slowed down to the velocity v.

It is the same as in the debates of the molecular velocity distribution in the atmosphere at the edge, the molecules have a certain distribution of energy. Those who arrive at the top have the same distribution of energy, since the slow ones did not arrive at all, and the fast ones were slowed down. The molecules that evaporate have the same distribution of energy as those inside a unique fact that is surprising.

Anyway, trying to argue so closely about our formula is useless because of other inaccuracies, like the likelihood of bouncing back rather than entering the liquid, and so on. But we do have a rough idea of the rate of evaporation and condensation and one can observe of course, that the density of vapor n varies in the same way as before, but now they have learned it in some detail rather than just as an arbitrary formula.

A greater knowledge helps us to examine other issues. For example, if they were to pump out the vapor at such a high rate that we extracted the vapor as quickly as it formed, if they had very good pumps and the liquid evaporated very slowly, how quickly would evaporation occur if we retained a liquid T? It is presumed that they have already calculated the equilibrium vapor density experimentally, so that one knows the number of molecules per unit volume are in equilibrium with the liquid at the given temperature.

At this point scientist want to know how quickly this will evaporate. While they have used only a rough analysis as far as the evaporation aspect of it is concerned, the amount of vapor molecules that arrived was not done so badly, apart from the unknown factor of the coefficient of reflection.

So, one can use the idea that, at equilibrium, the number that leaves are the same as the number that arrives. Yes, the vapor is being swept away, and so the molecules just fall out, but if the vapor is left alone, it would reach the density of equilibrium at which the number returning would exceed the number that are evaporating.

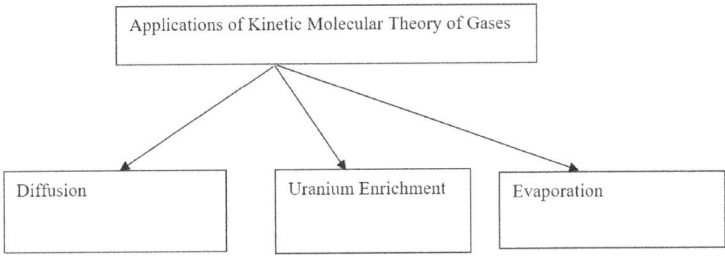

**Figure 4.6.** Application of kinetic molecular theory of gases.

Therefore, we can easily see that the amount that falls off the surface per second is equal to one minus the unknown reflection coefficient R times the number that would come down to the surface every second if the vapor were still there, because that is how many would match the evaporation at equilibrium:

Ne= nv(1−R) =(v(1−R)/Va) e−W/kT.

Obviously, the number of molecules that enter the liquid is equivalent to one.

## 4.7. CONCLUSION

The KMT of gases has been discovered in 19th century by James Clerk Maxwell and other scientist and in this chapter the properties have been explained. The Laws of KMT has been explained and the pressure of ideal gas has been useful in this context. There are many applications of KMT of gases and the most important use is in the uranium enrichment in the nuclear fission fuel production process. This is one of the most important process that has led to the discovery of process of manufacturing of nuclear fuel that is the raw material in the nuclear power plant for production of electricity. There are other applications also like diffusion and evaporation.

# REFERENCES

1.    Boudreaux, M., (2020). Properties of Gases. [ebook] p. 2. Available at: https://www.angelo.edu/faculty/kboudrea/index/Chapter_06–2.pdf (accessed on 13 March 2020).

2.    Brush, S., (n.d.). History of the Kinetic Theory of Gases. [ebook] p .2. Available at: http://www.terpconnect.umd.edu/~brush/pdf/ITALENC. pdf (accessed on 13 March 2020).

3.    Chemistry LibreTexts, (2019). 6.4: Kinetic Molecular Theory (Overview). [online] Available at: https://chem.libretexts. org/Bookshelves/General_Chemistry/Book%3A_Chem1_ (Lower)/06%3A_Properties_of_Gases/6.04%3A_Kinetic_Molecular_ Theory_(Overview) (accessed on 13 March 2020).

4.    Chemistry LibreTexts, (2019). Basics of Kinetic Molecular Theory. [online] Available at: https://chem.libretexts.org/Bookshelves/ Physical_and_Theoretical_Chemistry_Textbook_Maps/ Supplemental_Modules_(Physical_and_Theoretical_Chemistry)/ Physical_Properties_of_Matter/States_of_Matter/Properties_of_ Gases/Kinetic_Theory_of_Gases/Basics_of_Kinetic_Molecular_ Theory (accessed on 13 March 2020).

5.    Encyclopedia Britannica, (n.d.). Kinetic Theory of Gases | Physics. [online] Available at: https://www.britannica.com/science/kinetic-theory-of-gases (accessed on 13 March 2020).

6.    Feynmanlectures.caltech.edu. (n.d.). The Feynman Lectures on Physics Vol. I Ch. 42: Applications of Kinetic Theory. [online] Available at: https://www.feynmanlectures.caltech.edu/I_42.html (accessed on 13 March 2020).

7.    Grc.nasa.gov. (n.d.). Kinetic Theory of Gases. [online] Available at: https://www.grc.nasa.gov/www/k-12/airplane/kinth.html (accessed on 13 March 2020).

8.    Key, J., & Ball, D., (2020). Kinetic Molecular Theory of Gases. [online] Opentextbc.ca. Available at: https://opentextbc.ca/ introductorychemistry/chapter/kinetic-molecular-theory-of-gases-2/ (accessed on 13 March 2020).

9.    Kinetic Theory, (n.d.). [ebook] NCERT, p. 2. Available at: http://www. ncert.nic.in/NCERTS/l/keph205.pdf (accessed on 13 March 2020).

10.   Robinson, A., (2018). What is a Manometer? [online] Sciencing. Available at: https://sciencing.com/manometer-2718.html (accessed

on 13 March 2020).

11.  Semat, H., & Katz, R., (n.d.). Physics, Chapter 16: Kinetic Theory of Gases. [online] Digitalcommons.unl.edu. Available at: https://digitalcommons.unl.edu/cgi/viewcontent.cgi?article=1151&context=physicskatz (accessed on 13 March 2020).

12.  The Kinetic Theory of Gases, (n.d.). [ebook] p. 2. Available at: https://www.utsc.utoronto.ca/~quick/PHYA10S/LectureNotes/LN-17.pdf (accessed on 13 March 2020).

13.  Zimmerman, J. A., (2019). Kinetic Molecular Theory of Gases. [online] Available at: https://www.thoughtco.com/kinetic-theory-of-gases-2699426 (accessed on 13 March 2020).

# Thermochemistry of Chemical Substances

## CONTENTS

This chapter explains the basic significance of the chemical substances originating from the basic overview on elements and compounds as well as the physical and chemical properties of substances. A section that explains the significance of the chemical mixture and the thermochemistry of the homogenous mixture and heterogenous mixture is also created to distinguish mixtures from pure substances. This chapter also provide highlights on the phase transitions such as vaporization and condensation, boiling points, enthalpy of vaporization, melting, and freezing point, and addresses the process of sublimation and deposition, dissolution, and the formulation of the solutions.

# 5.1. INTRODUCTION

In the context of chemistry, a chemical substance is type of matter that has a constant composition of chemicals and their own unique or distinctive characteristic properties. It cannot be separated into different components without breaking their chemical bonds. Chemical substances could be in the form of solids, liquids, gases, and plasma. Alteration or modifications in the temperature or pressure can produce shift in the substance between the several types of various phases of matter.

An element is a type of chemical substance that is made up of specific kind of atom and in this way, that element cannot be broken down or altered with the help of chemical reaction into the various elements. All atoms that consist an element have the equal number of protons, nevertheless, they may have several numbers of electrons and neutrons.

A pure chemical compound is a chemical substance that is made of a specific set of molecules or ions that are chemically bonded. Two or more elements that are integrated into one substance with the help of a chemical reaction, for example water, creates a chemical compound. All compounds are basically a substance, but the vice-versa is not true.

A chemical compound can be either atoms that are bonded with each other in molecules or crystals in which the atoms, ions, or molecules creates a crystalline lattice. Generally, the compounds are made up of atoms of hydrogen and carbon that are known as organic compounds. On the other hand, all others are known as inorganic compounds. Compounds that are containing the bonds between the carbon and a metal are known as organometallic compounds.

Chemical substances that are, every so often, called 'pure' in order to set them apart from the mixture. A common instance of a chemical substance is the pure form of water (H2O). H2O always, has the same properties or characteristics and the same ratio of hydrogen to oxygen whether it is isolated from a river or made in a laboratory.

In addition to this, other chemical substances that are generally found in the pure form are diamond (a form of carbon), table salt (also known as sodium chloride), gold, refined sugar (sucrose). In simpler words, pure substances that are found in nature can be mixtures of chemical substances.

For instance, tap H2O can ne consisting of small amounts of dissolved sodium chloride (NaCl) and the compounds that are consisting of calcium, iron, and several numbers of other chemical substances as well. Pure distilled H2O is a substance, but sea H2O, as it is consisting of ions and several numbers of complicated molecules, is a mixture.

# 5.2. CHEMICAL MIXTURES

A mixture is system that is made up of two or more kinds of substances. It also refers to a physical integration or amalgamation of two or more substances in which the identities of the individual substances are reserved. Mixtures can be present in the form of solutions, colloids, alloys, and suspensions as well.

## 5.2.1. Heterogeneous Mixtures

A heterogeneous mixture is a mixture which is consisting of two or more substances (they could be elements or could be compounds), where the various types of components can be visually differentiated and easily separated with the help of physical or chemical means. Some examples of mixtures are:

1. mixtures of sand and H2O;
2. mixtures of sand and iron filings;
3. a conglomerate rock;
4. H2O and oil;
5. a salad;
6. trail mix; and
7. mixtures of gold powder and silver powder.

## 5.2.2. Homogenous Mixtures

A substance is homogeneous if its composition is identical wherever you sample it-it has uniform composition and properties throughout. Homogeneous is Latin for the same kind. If a substance is not homogeneous, it is said to be heterogeneous.

A homogenous mixture is a mixture of two or more chemical substances (they could be elements or could be compounds as well) that exhibits a single phase. In homogenous mixture, the several numbers of various compounds cannot be visually differentiated. The composition of the homogenous mixtures is constant. Every so often, separating the components of a homogenous mixture is more difficult task as compared to separating the components of a heterogenous mixture.

The main difference between homogenous and heterogenous mixture is its final phase. The combining substances were no longer distinguished in a homogenous mixture, which means separating them will require so much effort is not, a more serious method. On the other hand, heterogenous mixtures always distinguishes the reacting substances. Being able to see which is which, heterogeneous substances are easier to separate and many can be separated by physical and mechanical means.

With respect to practical terms, if the characteristics or the properties of interest is the same, irrespective of how much of the mixture is taken, the mixture is homogenous. The physical properties of the mixture, for example, its melting point, can fluctuate from those of its individual components. Some of the mixtures can be separated into their components with the help of the physical (that is thermal or mechanical) means.

## 5.3. ELEMENTS

A chemical element is a pure substance that is consisting of only one type of atom. Each atom has an atomic number. The atomic number of the atom represents the numbers of protons that are present in the nucleus of a single atom of that element. The periodic table of the element is ordered with the help of the atomic number in ascending order.

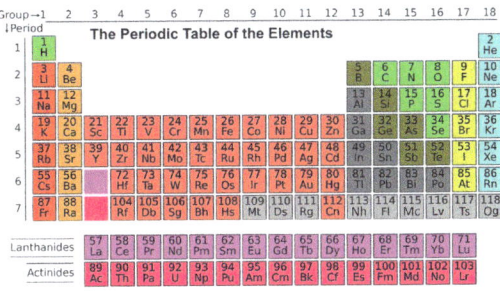

**Figure 5.1**. An illustration of the periodic table.

*Source: Image by Pixabay.*

The chemical elements are separated into the metals, the non-metals, and the metalloids. Conventionally, the metals are found on the left side of the periodic table and are:

1.    Every so often conductive to electricity;

2.    malleable in nature;

3.    Shiny; and

4.    Often shows magnetic properties as well.

Aluminium, iron, copper, gold, mercury, and lead are metals.

In contrast, non-metals are found on the right side of the periodic table (to the right of the staircase), are:

1.    Generally, they are not conductive in nature;

2.    Not malleable in nature;

3.    dull (not shiny); and

4.    Does not shows magnetic properties.

Examples of elemental non-metals is consisting of carbon and oxygen.

Metalloids have some characteristics or properties of metals along with some characteristics of non-metals. Silicon and arsenic are metalloids.

As of November 2011, a total of 118 elements have been found out (the most recently identified was ununseptium, in the year 2010). Of 118 known elements, only the first ninety-eight elements are known to the Earth's crust, while some are synthesized in the laboratory and a few of them can be made with the help of man-made nuclear reactions.

Out of 98 substances, only 80 of them are found on Earth or naturally occurring elements and they are found to be stable as well, which means they decay into the lighter elements over the period of time which is ranging from the fractions of seconds to the hundreds and thousands of years.

By far, only hydrogen and helium are the most abundant elements all across the universe. Nevertheless, with respect to mass, then the iron is most abundant element in the composition of the Earth. On the other hand, oxygen is the most common element in the layer that is the crust of the Earth.

## 5.4. COMPOUNDS

Pure substances are pure samples of the isolated elements that are uncommon in nature. The 98 naturally occurring elements that have been identified are in mineral samples from the crust of the Earth. Only a small portion of them can be found as recognizable, comparatively, pure minerals. From the more common of these kinds of "native elements" are gold, copper, sulfur, and silver.

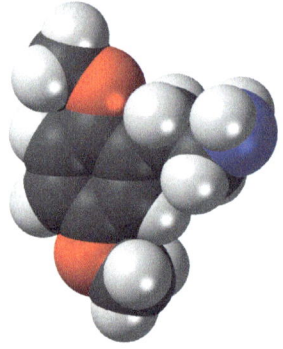

**Figure 5.2.** An illustration of the compound.

*Source: Image by Needpix.com.*

Carbon is an element that is commonly found in the form of graphite, coal, and diamonds. The noble gases (such as neon and the rest) and noble metals such as gold silver and mercury, can also be found in their pure, non-bonded forms in nature. Still, most of these kinds of elements are found in mixtures.

When two different types of elements are chemically amalgamated or combined, a chemical compound is formed. Most of the elements on the Earth combine with other elements in order to form a stable chemical compound, for example, sodium (Na) and chloride (Cl), which when allowed to react form table salt (that is NaCl). H2O is another example of a chemical compound. The component elements of a compound can be separated with the help of chemical reactions.

Chemical compounds have a distinct and a defined structure, which is consisting of a fixed ratio of the atoms which are held together in a defined spatial arrangement maintained by a chemical bond. Chemical compounds can be:

1. molecular compounds held together with the help of covalent bonds;

2. salts held together with the help of ionic bonds;

3. intermetallic compounds held together with the help of metallic bonds; and

4. complexes held together with the help of coordinate covalent bonds.

Pure chemical elements are not well thought out as a chemical compound, even if they are consisting of a diatomic or polyatomic molecule (are those molecules that are consisting of only multiple atoms of a single element, such as H2 or S8).

## 5.5. PROPERTIES OF MATTER

Each and every properties of characteristics of the matter, are either intensive or extensive and either physical as well as chemical. Extensive properties are those properties that rely on the amount of matter that are being measured, for example, mass, and volume. Intensive properties are those properties that rely on the amount of the matter, for example, density, and colour.

Both extensive as well as intensive properties are the physical properties, that means they can be measured without the alteration, modification or destruction of the chemical identity of the substance. For instance, the freezing point of a substance is a physical property: when H2O freezes, it's still made up of definite proportions of hydrogen and oxygen ; it is just in a different physical state.A chemical property, on the other hand, is a property of material that becomes evident through the course of the chemical reaction, that is, any property that can be developed only with the help of alteration, modification or destruction of the chemical identity of the substance. Chemical properties cannot be found out just by looking at or observing or touching the substance. The internal structure of the substance must be impacted for its chemical properties in order to be investigated or evaluated.

## 5.5.1. Physical Properties

Physical properties are the properties that can be evaluated or assesses or observed without altering the chemical nature of the substance. Some of the example of the physical properties are mentioned below:

1. colour (an intensive property);
2. density (an intensive property);
3. volume (an extensive property);
4. mass (an extensive property);
5. boiling point (an intensive property): the temperature at which a substance boil; and
6. melting point (an intensive property): the temperature at which a substance melt.

## 5.5.2. Chemical Properties

There are several numbers of example that are related to the chemical properties of a substance and the are:

1. Heat of combustion is the energy that is released when a compound goes through the complete combustion (burning) with oxygen. The symbol for the heat of combustion is $\Delta Hc$.
2. Chemical stability refers to whether a compound will react with $H2O$ or air (chemically stable substances will not react). Hydrolysis and oxidation are two types of reactions and are both chemical changes.
3. Flammability refers to whether a compound will burn when it comes near to the flame. Again, burning is a chemical reaction, which is commonly a high-temperature reaction in the presence of oxygen.
4. The preferred oxidation state is the lowest-energy oxidation state that a metal will undergo reactions in order to achieve (if another element is present to accept or donate electrons).

## 5.6. PHASE TRANSITIONS

The changes in the physical state, or phase transitions, can be introduced to a substance in a great number of approaches. As an example of the global significance, is the condensation process which allows rain to pour, melting of ice which cools the environment, evaporation of water which dries our wet clothes, and freezing of H2O turning it to ice, which we use to make beverages more refreshing.

These alterations and modifications of state are very important with respect to the H2O cycle of earth, along with several numbers of various natural phenomena as well as technological processes that are essential to the human life.

## 5.6.1. Vaporization and Condensation

When a liquid vaporizes in a closed container, the molecules of the gas cannot escape from the container. As these gas phase molecules move randomly about, they will occasionally collide with the surface of the condensed phase. On the other hand, in some cases, these types of collisions will result in the molecules re-entering the condensed phase. The alteration or modification from the gas phase to the liquid is known as condensation.

When the rate of the condensation becomes equal to the rate of vaporization, neither the amount of the liquid nor the amount of the vapor in the container changes. The vapor that are present in the container is then said to be in the state of equilibrium with the liquid. Keeping in mind that this is not a static circumstance, as the molecules are continually exchanged between the condensed and the gaseous phases.

As an example of the dynamic equilibrium, the status of a system in which the reciprocal processes (for instance, the vaporization and the condensation), that take place with same rates. The pressure that is exerted with the help of the vapor in the equilibrium with a liquid in a closed container at a given temperature is called the vapor pressure of the liquid (or the equilibrium vapor pressure).

The area of the surface of the liquid which is in contact with a vapor and the size of the vessel have no impacts on the vapor pressure, nevertheless, they do impact the time that is needed with the respect to the equilibrium to be accomplished.

The pressure of the vapor of a liquid can be measured with the help of placing a sample in a closed container, same as shown in the Figure 5.3. And also, with the application of the manometer in order to measure the pressure, the condensation will be better explained.

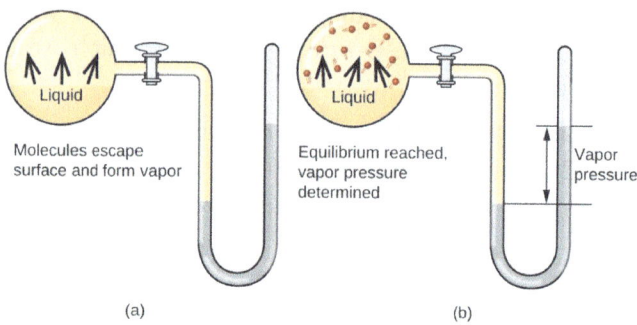

**Figure 5.3.** Vaporization and condensation in a closed container.

*Source: Image by Wikimedia Commons.*

In a closed container, dynamic equilibrium is reached when (a) the rate of molecules escaping from the liquid to become the gas (b) increases and eventually (c) equals the rate of gas molecules entering the liquid. When this equilibrium is reached, the vapor pressure of the gas is constant, although the vaporization and condensation processes continue

The chemical identities of the molecules that are present in a liquid allows us to find out the types (as well as the strengths) of the inter molecular attraction in the molecule. As an outcome, the several substances will exhibit various equilibrium vapor pressures.

Comparatively, the strong inter molecular attractive forces will provide assistance to delay the process of the vaporization along with the inclination to recapture the gas-phase molecules when they collide with the liquid surface, in a comparatively low pressure of the vapor.

The weak intermolecular attractions that shows less of a blockade with respect to the vaporization, and a decrement in the likelihood of the gas recapture, comparatively producing high pressure of the vapor. The following examples that shows the dependence of the vapor pressure on the intermolecular attractive forces.

Changes in temperature, also results in the change in pressure of the vapor or a liquid and the reason behind is  that is because the average kinetic energy (KE) of the molecules also changes. Recall that at any given temperature, the molecules of the substances that undergoes the change of KE, with a certain fraction of the molecules that are having appropriate amount of energy, overcome the inter molecular force (IMF). Some of this change in energy cause the change in behaver of the molecules, effecting changes in the phase. The evaporation of water for instance generated enough change in KE of the water molecules to orient themselves farther apart from each other, allowing them to move freely and therefore mix themselves homogenously with air molecules.

At a higher temperature, a greater fraction of the molecules has sufficient amount of energy in order to escape from the liquid, as shown in the Figure 5.4. The escape of more molecules per unit and the greater average speed of the molecules as it escapes, effect to higher pressure of the vapor.

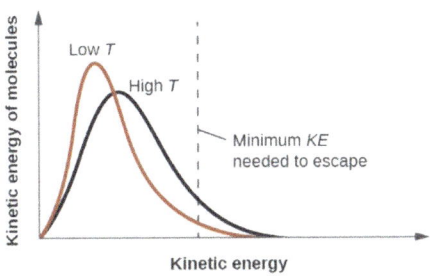

**Figure 5.4.** Temperature affects the distribution of kinetic energies for the molecules in a liquid.

*Source: Image by Wikimedia Commons.*

At higher temperatures, more molecules have the necessary KE, to escape from the liquid into the gas phase.

## 5.6.2. Boiling Points

When the pressure of the vapor elevated to a certain amount and reaches the pressure of external atmosphere, then the liquid starts boiling point. The boiling point of the liquid is the temperature at which its equilibrium vapor pressure is same as the pressure that is exerted on the liquid with the help of the gaseous surrounding. With respect to the liquids in open containers, it boils because it reaches the temperature where its pressure (pressure of the system) is equal to the atmospheric pressure (pressure of the surroundings).

The normal boiling point of a liquid is defined as its boiling point when the surrounding pressure is equal to 1 atmosphere or 1 atm (101.3 kPa). As shown in the Figure 5.5, the fluctuations in the pressure of the vapor with that of the temperature for several substances. The curves shows the dependence of the boiling point of the liquid with that of the surrounding pressure.

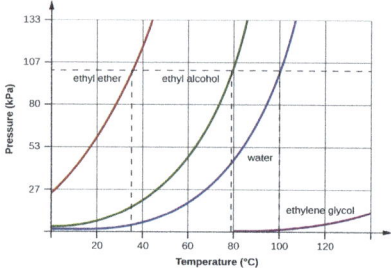

**Figure 5.5.** Graph representing boiling points of different liquids.

*Source: Image by Wikimedia Commons.*

The quantitative relation among the pressure of the vapor of any substance and the temperature of the substances is explained with the help of the Clausius-Clapeyron equation:

$$P=Ae-\Delta Hvap/RT \tag{5.1}$$

where, $\Delta Hvap$ is the enthalpy of the vaporization for the liquid, and R is the universal gas constant, and A is a constant that relies on the chemical identity of the substance. This equation is so often rearranged into its logarithm form in order to produce the linear equation:

$$ln*P= -(\Delta Hvap/RT) + ln*A \tag{5.2}$$

The above linear equation (5.2) can be expressed in a two-point format that is suitable or appropriate with respect to its use in several numbers of different computations, as shown in the equation (5.3) given below. If at specific temperature, let's assume T1, the pressure of the vapor is P1, and at the temperature T2, the pressure of the vapor is P2, then the corresponding linear equation can be written as mentioned below:

$$lnP1=-\Delta Hvap/RT1+lnA \tag{5.2.1}$$

$$lnP2=-\Delta Hvap/RT2+lnA \tag{5.2.2}$$

Since the constant, ln A, is the same, these two equations can be rearranged in order to isolate ln A and then set equation (5.2.1) and (5.2.2) equal to one another, can be written as:

$$lnP1+\Delta Hvap/RT1=lnP2+\Delta Hvap/RT2 \tag{5.3}$$

which can be integrated in order to get the below mentioned equation:

$$ln(P2/P1) = \Delta Hvap/R(1/T1-1/T2) \tag{5.4}$$

## *Example*

Calculate the vapor pressure of 1-butanol at 56 °C when its vapor pressure at 14.22 °C is 3.08 torr.

Given:

P1 = 3.08 torr                    T1 = 14.22 + 273.15 = 287.37 K

P2 = ? torr          T21 = 56 + 273.15 = 329.15 K

$\Delta$Hvap(1-butanol) = 43.29 kJ/mol

R = 0.008314 kJ/K·mol

Solution:

Using equation (5.4):

ln(P2/P1) =$\Delta$Hvap/R(1/T1−1/T2)

ln[3.08 torr/PT2,vap] = (43.29 kJ/mol/0.008314 kJ/K·mol)[1/287.37 K - 1/329.15 K]

ln[3.08 torr/PT2,vap] = 5206.88(4.42 x 10-4)

ln[3.08 torr/PT2,vap] = 2.30

PT2,vap/3.08 torr = 9.97

PT2,vap = 30.72 torr

## 5.6.3. Enthalpy of Vaporization

Vaporization is an endothermic process. The cooling effect can be evident when you leave a swimming pool or a shower. When the H2O on your skin evaporates, it removes heat from your skin and causes you to feel cold. The energy change associated with the vaporization process is the enthalpy of vaporization, $\Delta$Hvap. For example, the vaporization of H2O at standard temperature is represented by:

H2O(l)→H2O(g)                    $\Delta$Hvap=44.01 kJ/mol

Thus, the reverse of an endothermic process is an exothermic process and so, the condensation of a gas releases heat:

H2O(g)→H2O(l)                    $\Delta$Hcon= −$\Delta$Hvap= −44.01kJ/mol

## 5.6.4. Melting and Freezing

When a crystalline solid is  heated, the change in the average energy of the atoms, ions, or molecules and the solid corresponds with increases in temperature heating further the system. At some point o, the additional energy becomes large enough to partially overcome the forces that are holding the molecules or the ions (of the soli)d in their fixed positions, and then the solid starts transitioning to the liquid state, or melting.

Upon melting, the increment in the temperature stops, and the supply of heat remains constant until all of the solid is melted. Only after all of the solids are melted, the continued heating is then considered as increment in the temperature of the liquid (as shown in the Figure 5.6. below).

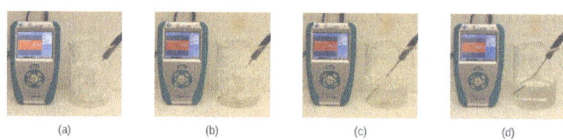

**Figure 5.6.** Temperature monitoring of the melting process (ice to liquid water). (a) The ice in the beaker has a temperature of −12.0°C. (b) After 10 minutes the ice has absorbed enough heat from the air to warm to 0°C. A small amount of ice has melted. (c) Thirty minutes later, the ice has absorbed more heat, but its temperature is still 0°C. The ice melts without changing its temperature. (d) Only after all the ice has melted does the heat absorbed cause the temperature to increase to 22.2°C.

If the supply of heat is halted through the course of melting and allocate the mixture of solid and the liquid in a perfectly insulated container, no additional heat can transit into or out of the container.

This is almost the situation with the mixture of ice and the H2O in an insulated material. There is no heat that can penetrates or escape from the system, and the mixture of the solid ice and liquid H2O remains at that condition for a long time.

In a mixture of liquid and solid at equilibrium, the reverse process of the melting is boiling which take place at same rate, and in this way, the quantities of the solid and the liquid remains constant. The temperature at which the liquid and the solid phase of a given substance are in the state of equilibrium is known as the melting point and the and the freezing point, respectively, (of the liquid).

Application of the one terms is normally dictated by the direction of the phase transition being well thought out, for instance, solid to liquid (is known as melting) or liquid to solid (is known as freezing).

The enthalpy of the fusion process and the melting point of a crystalline solid relies on the strength of the attractive forces between the units that are present in the crystal. Molecules that are having weak attractive forces from the crystals with low melting points. Crystals that are consisting of the particles with stronger attractive forces, will melt at higher temperatures.

The amount of the heat that is needed in order to alter or modify one mole of the substance from the solid state into the liquid state is the enthalpy of fusion or simply heat of fusion, written as ΔHfus of the substance. The enthalpy of the fusion of the ice is 6.0 kJ/mol at the temperature at 0°C or 273.15 K. Fusion or melting is an endothermic process, and the equation of the same can be written as:

H2O(s) →H2O(l)                    ΔHfus=6.01 kJ/mol

The reverse process, therefore known as freezing, is an exothermic process in which the enthalpy is altered or modified into −6.0 kJ/mol at 0°C:

H2O(l) →H2O(s)                    ΔHfrz= −ΔHfus= −6.01kJ/mol

## 5.7. SUBLIMATION AND DEPOSITION

Some solids have the have the ability to directly transform into its gaseous state, bypassing the liquid state, known as sublimination. At room temperature and at the standard pressure, a piece of dry ice (such as solid carbon dioxide ($CO_2$) sublimes, appearing to slowly disappear without ever forming any liquid.

Snow and ice sublime at temperature below the melting point of the H2O, a gradual process that can be speed up with the help of the winds and natural lowering of the atmospheric pressure at high amplitudes. When iodine in the solid form is heated, it sublimes and a vapor with vibrant purple colour is formed (as shown in the figure mentioned below).

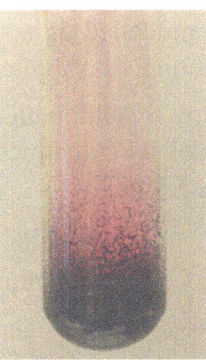

**Figure 5.7**. Sublimation of solid iodine when subject to heat.

In Figure 5.7, the sublimation of solid iodine at the bottom of the tube produces a purple gas that subsequently deposits as solid iodine on the colder part of the tube above. The reverse process sublimation is known as deposition, and is described as a process in which the gaseous substance condenses directly into its solid form. The formation of frost is an example of deposition.

Similar to the process of vaporization, the process of the sublimation necessitates an input of the energy in order to overcome or broken down the inter molecular bonds or atomic attractions. The enthalpy of the sublimation, which is denoted as $\Delta Hsub$, is the energy that is needed to transform one mole of substance from solid to its gaseous state. For instance, the sublimation of the $CO_2$ can be represented by the equation:

$CO_2(s) \rightarrow CO_2(g)$           $\Delta Hsub=26.1$ kJ/mol

By rule, the deposition of $CO_2$ which is the opposite of sublimation is written as:

$CO_2(g) \rightarrow CO_2(s)$           $\Delta Hdep= -\Delta Hsub= -26.1$kJ/mol

Considering the level to which the intermolecular bonds or attractions must be broken i to accomplish a given phase transition, transforming a solid into its liquid form necessitates that these kinds of bonds or attractions be only partially overcome or broken. In the transition to the gaseous state, those bonds, on the other hand, needs to be entirely broken. This results in the enthalpy of the fusion of that substance to be less than its enthalpy of vaporization.

This mechanism can also be used to drive an approximate association between the enthalpies of all phase changes with respect to the given substances. Even though, this process is not completely a precise explanation, sublimation may be suitably modelled as a series of two-step process of melting followed by vaporization to the application of the Hess Law.

In this context, the enthalpy of sublimation with respect to a substance can be estimated as the total of its enthalpies of fusion and enthalpies of vaporization, as shown Figure 5.8 below:

Solid→liquid           $\Delta Hfus$                    (5.5)

Liquid→gas            $\Delta Hvap$                   (5.6)

---

Solid→gas            $\Delta Hsub=\Delta Hfus+\Delta Hvap$           (5.7)

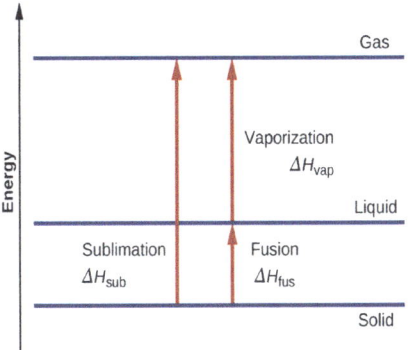

**Figure 5.8.** The sum of its enthalpy of fusion and enthalpy of vaporization is nearly equal to its enthalpy of sublimation.

*Source: Image by Wikimedia Commons.*

## 5.7.1. Heating and Cooling Curves

In the previous sections, the relationship between the amount of the heat of a substance, Q, and the temperature change that is associated with it can be written as ΔT, was introduced in the heat equation:

Q=mcΔT

where m is the mass and c is the specific heat of the substance corresponding to the change in temperature. The relationship applies to heating or cooling, but not phase change. When a substance is heated or cooled, it reaches a temperature corresponding to one of more of its phase transitions. In addition to this, the gain or loss of heat is the result of altering the inter molecular bonds or attractions.

On the other hand, in a substance is experiencing phase change, the temperature remains the same. Figure 5.9 illustrates the conventional heating curves of the different phase transformations of H2O.

Take for example heating a pot of H2O to boiling. A stove burner will supply the required amount of heat at a roughly constant rate, at the beginning, this heat facilitates the increase in the temperature of H2O. When the H2O reaches its boiling point, the temperature remains constant, for some time even the stove is still turned on. .

This same temperature is maintained by the H2O as long as it is boiling. If the setting of the burner is increased the amount of supplied to it was also increased, however, the temperature of H2O will not rise, however, since the amount of heat added is being consumed for the phase transformation.

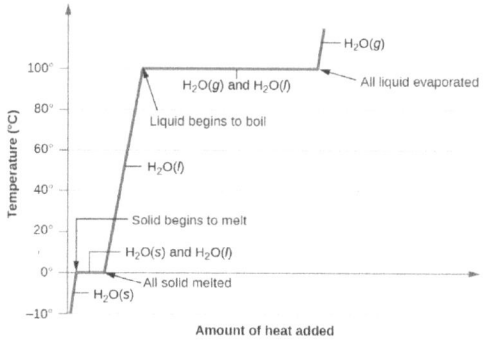

**Figure 5.9.** Heating Curves

A typical heating curve for a substance depicts changes in temperature that result as the substance absorbs increasing amounts of heat. Plateaus in the curve (regions of constant temperature) are exhibited when the substance undergoes phase transitions.

## 5.8. THE DISSOLUTION PROCESS

In a solution, the ratio of the mixing components was never equal. This is because the proportion of the solvent is always greater than that of the solute.

. Sugar is a covalent solid which is made of molecules of sucrose, $C_{12}H_{22}O_{11}$. When this compound dissolves in H2O, the molecules of the compound become uniformly distributed between the molecules of the H2O. And the equation can be written as:

$$C_{12}H_{22}O_{11}(s) \rightarrow C_{12}H_{22}O_{11}(aq)$$

The subscript "aq" in the equation depicts that the molecules of the sucrose are individually dispersed all across the solvent in the solution.. Although, molecules of the sucrose are much heavier as compared to the molecules of the H2O, they remain dispersed all across the solution and gravity is unable to "settle them down" over the interval of time.

Potassium dichromate, $(K_2Cr_2O_7)$, which is an ionic compound composed of colourless potassium ions, $K^+$, and dichromate ions which are orange to yellow in colour, $(Cr_2O_7^{2-})$. When a small amount of the solid potassium dichromate is added to H2O, the compound dissolves and separates in to ions of potassium and dichromate that are uniformly distributed all across the mixture (as shown in the figure that is mentioned below), as indicated in the below mentioned equation:

K2Cr2O7(s)→2K+(aq)+Cr2O7(aq)

**Figure 5.10.** Potassium Dichromate Solution

When (left) K2Cr2O7 is mixed with (middle) H2O, it forms (right) a homogeneous orange solution. H2O is used often as a solvent in many solutions because it of its ability to dissolve a wide range of solutes. Nevertheless, almost any kind of gas, solid or liquid can play as a solvent. Several numbers of alloys are solid solutions of one metal that dissolved with other, as an example, coins consisting of nickel that is dissolved in copper.

Air is a gaseous solution, it is a homogenous mixture of oxygen, nitrogen along with the several numbers of other gases as well. Oxygen (in the form of gas), alcohol (in the form liquid) all dissolve in H2O (in the form of liquid) in order to form a homogeneous liquid solution. Table 5.1. gives an example of several numbers of various solutions and the phases of the solvents and solutes.

**Table 5.1.** Different Types of Solutions

| Solution | Solute | Solvent |
|---|---|---|
| Air | $O_2$ (g) | $N_2$ (g) |
| Soft Drinks | $CO_2$ (g) | $H_2O$ (l) |
| Hydrogen in Palladium | $H_2$ (g) | Pd (s) |
| Rubbing Alcohol | $H_2O$ (l) | $C_3H_8O$(l) (2-propanol) |
| Saltwater | NaCl (s) | $H_2O$ (l) |
| Brass | Zn (s) | Cu (s) |

*Source: Table by Courses.Lumenlearning.*

Solutions exhibit these defining traits:

1.    They are homogeneous in nature; that is, after a solution is mixed, it has the same composition at all points all across (its composition is uniform).

2.    The physical state of a solution, can be solid, liquid, or gas. And, it is generally the same as that of the solvent, as demonstrated with the help of the examples in Table 5.1.

3.    The components of a solution are dispersed on a molecular scale; that is, they are consisting of a mixture of separated molecules, atoms, and/or ions.

4.    The dissolved solute in a homogeneous solution will not settle out or separate from the solvent.

5.    The composition of a solution, or the concentrations of its components, can be fluctuated continuously, within the limits.

## 5.8.1. The Formation of Solutions

The formation of a solution is an example of a spontaneous process, a process that basically arises under certain conditions without much need of energy importing from external source. Sometimes there is need of stirring a mixture in order to speed up the process of dissolution, but it is not a compulsion; a homogeneous solution would form automatically if one wait for a longer period of time.

The topic of spontaneity plays very important role in a way to have an understanding about chemical thermodynamics and is explained more extensively in later sections. For purposes of this chapter's discussion, it is utmost importance to take into consideration two criteria that favour, but do not assure, the spontaneous formation of a solution:

1.    A reduction is shown in the context of the internal energy of the system (an exothermic change, which is, as an outcome, release of energy in the form of heat or light)

2.    an increment of disorder in the system (which shows a considerable increment with respect to the entropy of the system, as this will discuss later in topics of the chapter on thermodynamics of chemical substances)

Through the course of the process of dissolution, it is every so often seen that a modification or alteration in the internal energy generally takes place, if not always, primarily because of the fact that heat often evolved absorbed.

When there is formation of a solution, there is always an increment in the disorder.

When the strong point of the intermolecular forces of attraction amid solvent and solute species in a solution are no different as compared to those present in the separated components, the solution is created with no accompanying energy change.

This type of solution is well thought out as an ideal solution. A combination of ideal gases (or gases such as argon and helium, which closely associate ideal behaviour) is an example of an ideal solution, since the entities containing these gases witness no substantial intermolecular attractions.

When containers, that are made of argon and helium are connected with each other, the gases mix spontaneously because of diffusion and form a solution (as shown in the Figure 5.11). The build-up of this solution undoubtedly results in an increment in disorder, since the argon and helium atoms occupy a volume two times the ones which is occupied before mixing.

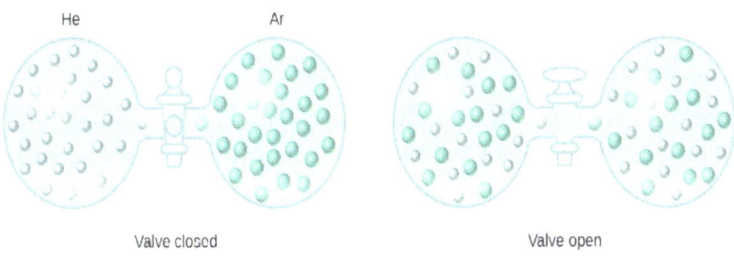

**Figure 5.11.** Diffusion and solution formation in gases.

Samples of helium and argon spontaneously mix to give a solution in which the disorder of the atoms of the two gases is increased.

There may be formation of ideal solutions when structurally similar liquids are amalgamated together. For instance, mixtures of the alcohols ethanol (C2H5OH) and methanol (CH3OH) create ideal solutions, in a similar way as done by the hydrocarbon pentane, C5H12, and hexane, C6H14.

Placing ethanol, and methanol, or hexane and pentane, in the bulbs shown in Figure that is mentioned above will result in the same diffusion and successive mixing of these liquids as is noticed for the Ar and He gases (although at a much slower rate), producing solutions with no much change in energy.

Dissimilar as a combination of gases, nevertheless, the mechanisms of these liquid-liquid solutions do, indeed, witness an intermolecular attractive force. Even though, as the molecules of the two substances that are being mixed are mechanically very comparable with each other, the intermolecular attractive forces between similar and dissimilar molecules are fundamentally the same, and the dissolution process, therefore, does not entail any appreciable decrease or increase in energy.

The examples as discussed above shows how diffusion alone can act a driving force to cause the spontaneous creation of a solution. In majority of the instances, the relative magnitudes of intermolecular forces of attraction between solvent and solute species may impede or obstruct the dissolution process.

It is worth noticing that there are three types of intermolecular attractive forces which play crucial role in the dissolution process: solvent-solvent, solute-solute, and solute-solvent. As explained in the Figure 5.12, the construction of a solution may be conceived as a stepwise process in which there is consumption of energy to prevent solvent-solvent and solute-solute attractions (endothermic processes) and free when there is establishment of solute-solvent attractions (an exothermic process referred to as solvation).

The relative magnitudes of the energy variations related with these stepwise processes evaluate whether the dissolution process will overall gain or losses energy. On the other hand, in some scenarios, there is no formation of solution as the energy needed to separate solvent and solute species is far larger comparatively with the energy released with the help of the solvation.

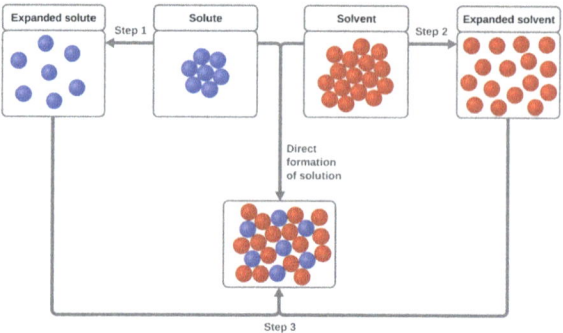

**Figure 5.12.** Schematic representation of dissolution.

*Source: Image by Wikimedia Commons.*

The schematic representation of dissolution shows a stepwise process involving the endothermic separation of solute and solvent species (Steps 1 and 2) and exothermic solvation (Step 3).

For instance, cooking oils and H2O will not blend to a substantial extent which resulted in yielding solutions (as shown in the figure that is mentioned above). Hydrogen bonding is the foremost important intermolecular attractive force exists in liquid H2O; the nonpolar hydrocarbon molecules of cooking oils have no potential of hydrogen bonding, instead being present together by dispersion forces.

Ensuring the formation of an oil- H2O solution, which requires preventing the very strong hydrogen bonding in H2O, as well as the imperative strong dispersion forces between the comparatively greater oil molecules. And, since the nonpolar oil molecules and polar H2O molecules would not witness very sturdy intermolecular attraction, very little amount of energy would be released by solvation.

On the other hand, a mixture of H2O and ethanol will mix in any proportions in order to produce a solution. In this scenario, both substances possess the ability of hydrogen bonding, and thus the solvation process is adequately exothermic to reimburse for the endothermic separations of solvent and solute molecules.

As it has been observed in the initial part of this chapter, it is always favourable to have formation of spontaneous solution, but not certain, by the processes of exothermic dissolution. While it is generally seen that majority of the soluble compounds do, certainly, dissolve with the release of heat, some dissolve endothermically.

One of the perfect examples of this is ammonium nitrate (NH4NO3). It is primarily used for the production of instant cold packs in order to treat injuries like the one described in the Figure 5. A thin-walled plastic bag of H2O is enclosed into a larger bag with solid NH4NO3.

Suppose if there is any breakage in the smaller bag, a solution of NH4NO3 forms, which started absorbing heat from the surroundings (the damaged area to which the pack is applied) and providing a cold compress which reduce the extent of swelling.

Endothermic dissolutions such as this one requires a large amount of energy input in order to separate the solute species than is recovered when the solutes are solvated, but they are natural nonetheless because of surge in disorder that follows creation of the solution.

## 5.9. CONCLUSION

In the conclusion of the chapter, the energy is the capacity to perform a work (applying a force in order to move a matter). KE is the energy of the moving objects, while potential energy refers to the energy relative to the object's position, condition, or composition. When energy is being converted from one to the other form, energy is neither be created nor be destroyed (according to the law of conservation of energy or according to the first law of thermodynamics).

Matter has thermal energy and the reason behind this is the relationship of temperature to the average KE of the molecules. Heat is the energy that is transferred between the objects at several temperatures; it flows from high temperature to the low temperature. Chemical as well as physical processes can absorb heat (that is endothermic process) or release heat (that is endothermic processes).The SI unit of the energy, work, and heat is the joule (J). Specific heat and heat capacity are measures of the energy that is required in order to modify the temperature of the substance or object. The amount of heat that is being absorbed or released with the help of a substance directly relies on the type of substance, the temperature change, and its mass it undergoes.

If a chemical change is done at a constant pressure and the only work done is by expansion or contraction, Q for the change is known as the enthalpy change with the symbol, symbol $\Delta H$, or $\Delta H298°$ for reactions that take place under the standard state settings.The value of the enthalpy $\Delta H$ for a reaction in one direction is equal in magnitude, but opposite in sign, to $\Delta H$ for the reaction in the opposite direction, and $\Delta H$ is directly related to the quantity of the reactants and the products. Example of the enthalpy changes which is consisting of enthalpy of combustion, enthalpy of fusion, enthalpy of vaporization, and standard enthalpy of formation.

The standard enthalpy of formation, that is denoted by $\Delta Hf°$, is the enthalpy change that is associated with the formation of one mole of a substance from the elements in their most stable states at one bar (which is the standard state). Several numbers of the processes are carried out at 298.15 K.

If the enthalpies of the formation are available for the reactant and the products of a reaction, the change in the enthalpy can be evaluated with the help of Hess's law. According to the Hess's law, if a process can be written as the sum of the several stepwise processes, the enthalpy change of the total process equals the sum of the enthalpy changes of the various steps.

# REFERENCES

1.    Bichowsky, F., & Rossini, F., (2003). Thermochemistry of the Chemical Substances. 3rd ed. [Norwich, N.Y.]: Knovel.

2.    Chapter 5 Thermochemistry, (2014). [ebook] US Energy Information Administration. Available at: https://web.ung.edu/media/chemistry/Chapter5/Chapter5-Thermochemistry.pdf (accessed on 13 March 2020).

3.    Lumen Learning, (n.d.). Elements and Compounds | Introduction to Chemistry. [online] Courses.lumenlearning.com. Available at: https://courses.lumenlearning.com/introchem/chapter/elements-and-compounds/ (accessed on 13 March 2020).

4.    Lumen Learning, (n.d.). Phase Transitions | Chemistry for Majors. [online] Courses.lumenlearning.com. Available at: https://courses.lumenlearning.com/chemistryformajors/chapter/phase-transitions-2/ (accessed on 13 March 2020).

5.    Lumen Learning, (n.d.). Physical and Chemical Properties of Matter | Introduction to Chemistry. [online] Courses.lumenlearning.com. Available at: https://courses.lumenlearning.com/introchem/chapter/physical-and-chemical-properties-of-matter/ (accessed on 13 March 2020).

6.    Lumen Learning, (n.d.). Substances and Mixtures | Introduction to Chemistry. [online] Courses.lumenlearning.com. Available at: https://courses.lumenlearning.com/introchem/chapter/substances-and-mixtures/ (accessed on 13 March 2020).

7.    Lumen Learning, (n.d.). The Dissolution Process | Chemistry for Majors. [online] Courses.lumenlearning.com. Available at: https://courses.lumenlearning.com/chemistryformajors/chapter/the-dissolution-process-2/ (accessed on 13 March 2020).

# Basic Thermochemistry for Material Processing

## CONTENTS

This chapter introduces various processes or reactions of thermochemistry that are used in material processing. There are a number of reactions of thermochemistry that are used in the processing of various materials, right from producing their monomers to manufacturing the finished product. This is very important to understand that Energy plays an important role in in the entire procedure of material processing. This chapter also explains the importance of maintaining the phase equilibrium.

The importance of vapor-liquid equilibrium is also explained in this chapter. Following the explanation of phase equilibrium through phase diagrams, this chapter introduces the various sources of energy such as the batteries and the fuel cells. There are two types of batteries that are used in the physical world. Both Primary batteries and secondary batteries are discussed in this chapter.

Further, this chapter explains the applications of electrochemistry and explains the various procedures of material processing that take place in the industrial setting. Lastly, this chapter discusses the phenomenon of corrosion and how it can be prevented with the use of thermochemistry in the procedures of material processing.

## 6.1. INTRODUCTION

Chemical reactions, such as those kinds of reactions that takes place when an individual light a match, includes the changes in energy as well as in matter. It has been observed that societies at all levels of development could not operate without the energy that is released by chemical reactions. In the year 2012, it has been observed that about 85% of Unites States energy consumption came from the combustion of petroleum products, wood, coal, and garbage.

This energy is used in many things. It is used to produce electricity (38%); to transport food, manufactured goods, raw materials, and people (27%); for industrial production (21%). In addition to, it is also used to heat and power homes and businesses (10%). On the other hand, these combustion reactions also help in meeting the essential energy needs, they are also recognized by the majority of the scientific community as a main contributor to global climate change.

The useful forms of energy are also available from a variety of chemical reactions other than just combustion. For instance, the energy that is produced by the batteries in a cell phone, flashlight or car, are the result of a chemical reactions. This chapter introduces many of the basic ideas that are required to discover the relationships that is between chemical changes and energy, with the emphasis on thermal energy.

All the chemical changes and their accompanying changes in energy are essential parts of the everyday world. The macronutrients that are present in the food (that is fats, proteins, and carbohydrates) go through metabolic reactions. It provides the energy that keeps the bodies functioning.

A variety of fuels (that is natural gas, gasoline, and coal) are burned in order to produce energy for several purposes. These are heating, transportation, and the generation of electricity. Industrial chemical reactions make use of enormous amounts of energy for the production of raw materials (like iron and aluminium). Then, the energy is used to manufacture those raw materials into valuable products. These products include skyscrapers, cars, and bridges.

It has been observed that more than 90% of the energy that an individual use comes originally from the sun. The sun provide energy to the earth every day. The sun provides the earth with almost 10,000 times the amount of energy that is required to meet all of the energy needs of the world for that day.

The main challenge is to find such kind of ways that can be used to convert as well as store incoming solar energy so that this energy can be used in reactions or chemical processes that are both convenient as well as non-polluting.

It has to be noted that all the plants and many bacteria capture solar energy through the process of photosynthesis. The energy that is stored in plants is released by the burning of wood or plant products like ethanol.

There are several uses of the energy. An individual can also make use of the energy to fuel their bodies by eating food that comes directly from plants or from those kinds of animals that got their energy by the consumption of plants. Solar energy is also released by burning coal and petroleum: These fuels are fossilized plant and animal matter.Materials processing includes a complicated series of thermal, chemical, and physical processes that make a starting material, create the shape of the material, retain that shape, and then refine the structure as well as shape. The main objective of materials processing is to develop the structural features (for example, microstructure, size, crystal structure, and shape).These structures are required for the product to perform its intended application in a better way. Materials processing is vital to the field of materials science and engineering. In addition to it, it is a vital step in manufacturing.

Thermochemistry is also used in material processing.

The conversion of the starting material to the final product takes place in the following three steps:

- Preparation of the starting material,
- Processing operation, and
- Post-processing operation(s).

The processing operations can be further divided into five categories. These categories are based on the state of matter that is most important to the process: solid, melt, dispersion or solution, powder, and vapor.

It has been observed that ceramics, polymers, and metals are formed by operations in each of the categories so that common scientific as well as engineering principles can be understood and used to several kinds of materials.

## 6.2. KEY TERMS IN MATERIAL PROCESSING

### 6.2.1. Energy

> The common forms of energy consist of potential energy and kinetic energy. Joule (J) is the SI unit of energy. Work is said to be done when a force causes an object to move.

Work = force × distance

### 6.2.2. Calorimetry

Measuring Quantities of Heat—Calorimetry is defined as the process of measuring quantities of heat. On the other hand, a calorimeter is defined as the device that is used for the procedure. The actual measurements are of temperature changes and masses.

The other data that is required are heat capacities and/or specific heats. The heat capacity (C) of a system is defined as the quantity of heat that is required to change the temperature of the system by 1°C (or to change it by 1 K):

The specific heat is the heat capacity of a 1-g sample of the substance. On the other hand, the molar heat capacity is the heat capacity of one mole of substance:

$$\text{Specific heat} = \frac{\text{heat capacity}}{\text{mass}} = \frac{C}{m} = \frac{q}{m \times \Delta T}$$

There are many reactions that can be carried out under the constant pressure of the atmosphere in a plastic-foam calorimeter. The results that are obtained from the reactions are qp or $\Delta H$ values. On the other hand, there are combustion reactions. These reactions are usually carried out under constant-volume conditions in a bomb calorimeter. The results that are obtained are qv or $\Delta U$ values; $\Delta H$ values can be calculated from $\Delta U$ values when it is required.

### 6.2.3. Hess's Law of Constant Heat Summation

The value of $\Delta H$ is the same whether a reaction is carried out in one step or it is carried out in several steps. This is because enthalpy is a state function. The equation for a reaction is written as several steps, and Hess's law is used to evaluate $\Delta H$. This is evaluated by summing the enthalpy changes of all the steps.

## 6.3. PHASE EQUILIBRIUM

Phase equilibrium is defined as the study of the equilibrium which exists between or inside different states of matter namely liquid, solid, and gas. Equilibrium is defined as a stage when chemical potential of any of the component that is present in the system stays steady with time.

Phase is defined as a region where all the intermolecular interaction is spatially uniform. In other words, physical, and chemical properties of the system are similar all through the region. A component can exist in two different phases like allotropes of an element within the same state. In addition to it, two immiscible compounds that are in same liquid state can exist in two phases.

It has been observed that phase equilibrium has wide range of applications in industries. These applications include lowering of freezing point of H2O by dissolving salt (brine), production of different allotropes of carbon, purification of components by distillation, use of emulsions in production of food, pharmaceutical industry, etc.

Solid-solid phase equilibrium has a special place in the field of metallurgy. In this field, it is used to make alloys of different physical as well as chemical properties. For example, melting point of alloys of copper and silver is lower as compared to the melting point of either copper or silver.

## 6.3.1. Phase Diagrams

Phase diagrams are used in order to understand the relationship that is between different phases. Generally, these are represented as the change in the phase of a system as a function of pressure, temperature or composition of the components in a system.

It has been observed that the system exists in a phase where Gibbs free energy of the system is least.

Also, it has been found that at equilibrium, pressure, temperature, and chemical potential of constituent component molecules in the system have to be similar throughout all the phases. Figure 6.1 provides a general schematic of phase diagram of a single component system (Lue, 2009).

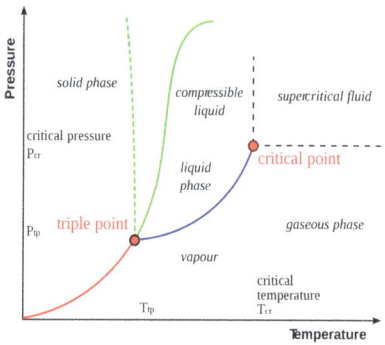

**Figure 6.1.** An illustration of phase diagram of one component system.

*Source: Image by Wikimedia Commons.*

The curves that are shown in the figure indicates the coexistence of two phases. Melting curve is defined as the curve in the phase diagram along which the solid as well as liquid phase of a system are in equilibrium.

Liquid as well as gas phase of a system stay in equilibrium along with the vaporization curve of the system. On the other hand, sublimation curve indicates the equilibrium stage that is between solid and gas phase of the system.

Triple point is defined as the point on the graph where all the three states that is liquid, gas, solid coexist and also, it is unique for every single component. If $\alpha$ and $\beta$ are any of the two phases in which a component can exist, by the use of first and second law of thermodynamics, the slope of any of the curves in Figure 6.1 can be represented by:

$$\frac{dP}{dT} = \left(\frac{\Delta S}{\Delta V}\right)_{\alpha \to \beta}$$
(6.1)

This equation is referred as Clausius Clapeyron equation. In equilibrium, vapor pressure means the total number of molecules evaporating is same to the number of molecules that are condensing on the surface of H2O.

This equation (1) can be written as

$$\frac{dP}{dT} = \frac{1}{T}\left(\frac{\Delta H}{\Delta V}\right)_{\alpha \to \beta}$$
6.2)

This equation is referred as Clausius Clapeyron equation. In equilibrium, vapor pressure means the total number of molecules evaporating is same to the number of molecules that are condensing on the surface of H2O.

In accordance to Gibbs phase rule regarding multi-component systems, the number of intensive degrees of freedom in a non-reactive multi-component system, F at equilibrium is given by

$$F = C - n + 2$$
(6.3)

In this reaction, C is the number of non-reactive components that are present in a system and n is the total number of phases. For instance, for two-component as well as two-phase systems, there are two intensive degrees of freedom which are names as either pressure, temperature or mole fraction.

In other terms, it has been found that in case of two-component and two-phase system at equilibrium, there are just two intensive variables that are required to uniquely determine the thermodynamic state of system.

There are different types of equilibrium that are examined in detail. These equilibriums include liquid-vapor equilibrium, solid-liquid equilibrium, liquid-liquid equilibrium, solid-solid equilibrium (alloys or allotropic forms), etc.

There are various causes of enhanced complexity of phase diagrams. Some of these causes are increase in number of components, presence of surfactants, chemical reactions, and deviation from ideal behaviour. In the below topics, a special case of vapor-liquid equilibrium is discussed which is commonly used in distilleries that is, ethanol-H2O mixtures.

## 6.3.2. Vapor-Liquid Equilibrium

Raoult's law says that in an ideal solution, the partial vapor pressure that is exerted by a component is the product of its mole fraction as well as vapor pressure of pure component. Thus, total vapor pressure that is exerted by an ideal solution with k components, P is shown below:

$$P = \sum_{i=1}^{k} p_i x_i$$

(6.4)

In this equation, xi is the respective mole fraction and pi is the vapor pressure of the pure component. For example, if in a brine solution (that is a two-component mixture), one component (that is common salt) has negligible vapor at the specified temperature then the vapor pressure of the mixture is simply product of the mole fraction of H2O as well as vapor pressure of H2O at specified temperature.

Raoult' law is just valid in case of ideal solutions. This assume no intermolecular interaction between different kind of components. On the other hand, it has been observed that in most of the cases that are dealt with in industries there are hardly any ideal solutions. Modified Raoult's law is applied for non-ideal solutions (J. M. Smith, 2005); accordingly, total pressure exerted, P is expressed as

$$P = \sum_{i}^{k} p_i \gamma_i x_i$$

(6.5)

In this equation, yi is the activity coefficient for the ith component.

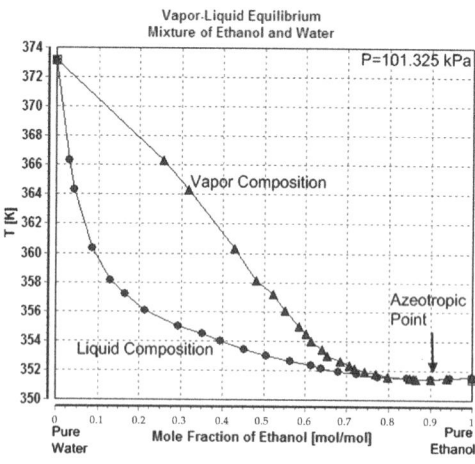

**Figure 6.2.** Vapor liquid equilibrium for ethanol water system.

*Source: Image by Wikimedia Commons.*

In Figure 6.2, the phase diagram for ethanol-H2O mixtures is shown. It has been observed that similar graphs are used in distilleries. These are used in order to calculate the desired temperatures for specific outlet compositions. It also includes a vapor curve and a liquid curve. This curve is used in order to evaluate the vapor or the liquid compositions.

The vapor curve is also known as dew point curve. On the other hand, the liquid curve is also known as bubble point curve. Interestingly, it has been observed that both these curves meet each other at their lowest point. This suggests that if an ethanol-H2O mixture is boiled further than it will have the same composition when it is in the vapor phase.

In this way, the maximum concentration of ethanol that is obtained by simple process of distillation in an ethanol-H2O mixture would be 95.5%. On the phase diagram, these mixtures which have the lowest/highest boiling point and also, they cannot be purified further. These are known as Azeotropes.

Azeotropes have higher or lower boiling point as compared to either of its constituent components. It depends on if the mixtures have positive deviation or negative deviation from Roult's law, respectively.

Triangular graphs are used with three end points of triangle. This represents the three pure components (phases) in case of three-component or three-phase systems.

# 6.4. INTRODUCTION TO ELECTROCHEMISTRY

Electrochemistry deals with all the chemical reactions that produce electricity and the changes that are linked with the passage of electrical current through matter. There is the transfer of electron in the reaction, and therefore they are oxidation-reduction (or redox) reactions.

In simple terms, electrochemistry is defined as the study of chemical processes that cause the electrons in the reaction to move. This movement of electrons is known as electricity. Electricity can be generated by movements of electrons from one element to the other element in a reaction. This is known as an oxidation-reduction ("redox") reaction.

It has been observed that there are many metals that may be purified or electroplated by the use of electrochemical methods. There are several devices like smartphones, automobiles, watches, electronic tablets, pacemakers, and many others that make use of batteries for power. Batteries make use of chemical reactions that generate electricity spontaneously and that can be transformed into some important and useful work. It has been observed that all electrochemical systems include the transfer of electrons in a reacting system. On the other hand, in many systems, the reactions take place in a region which is known as the cell. Here the transfer of electrons occurs at electrodes.

## 6.4.1. Balancing Oxidation-Reduction Reactions

Electricity refers to several phenomena that are linked with the presence as well as flow of electric charge. Electricity comprises such diverse things as static electricity, lightning, the current that is produced by a battery as it discharges, and many other influences on the day to day lives.

Electric current is the movement or flow of the charge. The charge is carried by electrons or ions. The elementary unit of charge is defined as the charge of a proton. This is equivalent in magnitude to the charge of an electron. Coulomb (C) is the SI unit of charge and the charge of a proton is $1.602 \times 10^{-19}$ C. Electric field is generated by the presence of an electric charge.

Electric current is defined as the rate of flow of electric charge. Ampere (A) is the SI unit for electrical current. It is a flow rate of 1 coulomb of charge for each second (1A=1C/s).

Electric circuit is defined as the path in which an electric current flow. It has been observed that in most of the chemical systems, it is required to maintain a closed path for electric current to flow.

The flow of charge is generated by an electrical potential difference, or potential, between two points in the circuit. Electrical potential is the ability of the electric field to do work on the charge. Volt (V) is the SI unit of electrical potential. 1 joule (J) of energy is gained or lost when 1 coulomb of electric charge moves through a potential difference of 1 volt.

Electrochemistry studies about the oxidation-reduction reactions. Oxidation means the loss of electrons and reduction means the gain of electrons. The reactions discussed tended to be rather simple, and conservation of mass (atom counting by type) and deriving a correctly balanced chemical equation were relatively simple.

The use of half-reactions is essential partly for balancing those reactions that are more complicated and also it is important because many aspects of electrochemistry can be discussed easily in terms of half-reactions.

There are different methods that are used for balancing these reactions. However, it has been observed that there are no good alternatives to half-reactions for discussing what is going on in many systems. In the half-reaction method, oxidation-reduction reactions are split into their oxidation "half" and reduction "half" to make the overall equation easier and understandable.

It has been observed that electrochemical reactions often occur in solutions. This could be basic, acidic, or neutral. The nature of the solution may be important at the time of balancing oxidation-reduction reactions. It is quite evident in an actual problem. Unbalanced oxidation-reduction reaction in acidic solution is illustrated below:

$$MnO_4^-(aq) + Fe^{2+}(aq) \longrightarrow Mn^{2+}(aq) + Fe^{3+}(aq)$$

This can be started by collecting all the species that we have so far into an unbalanced oxidation half-reaction and an unbalanced reduction half-reaction. It has been observed that every single half-reaction involves the same element in two different oxidation states. For example, The $Fe^{2+}$ has lost an electron in order to become $Fe^{3+}$. Thus, the iron undergoes the process of oxidation. Here, the reduction is not as obvious. However, the manganese changed from $Mn^{7+}$ to $Mn^{2+}$ by gaining five electrons.

Oxidation (unbalanced): $Fe^{2+}(aq) \longrightarrow Fe^{3+}(aq)$

reduction (unbalanced): $MnO_4^-(aq) \longrightarrow Mn^{2+}(aq)$

There is the presence of hydrogen ions present in acidic solution. Often, these are useful in balancing half-reactions. It may be required to make use of the hydrogen ions directly or it can also be used as a reactant that may react with oxygen in order to generate $H_2O$.

It has to be noted that hydrogen ions are very important in acidic solutions. In these solutions, the reactants or products contain hydrogen and/or oxygen. In the example that is given below, the oxidation half-reaction contains neither hydrogen nor oxygen, so hydrogen ions are not required to be balanced. However, oxygen gets necessarily involved in reduction half-reaction. It is also required to make use of hydrogen ions in order to convert this oxygen to $H_2O$.

charge not balanced: $MnO_4^-(aq) + 8H^+(aq) \longrightarrow Mn^{2+}(aq) + 4H_2O(l)$

On the other hand, in basic solution, the situation is different. This is because the concentration of hydrogen ion is lower, and the concentration of hydroxide ion is higher. After understanding the above equation, the next step is to examine how basic solutions are different from acidic solutions. A neutral solution may be treated as basic or acidic but treating the neutral solution as acidic is generally easier.

## 6.4.2. Galvanic Cells

Galvanic cells are also called voltaic cells. These are defined as electrochemical cells in which electrical energy is produced by spontaneous oxidation-reduction reactions. It is often convenient to separate the oxidation-reduction reactions into half-reactions in writing the equation. This is done to facilitate balancing the overall equation and also, to highlight the actual chemical transformations.

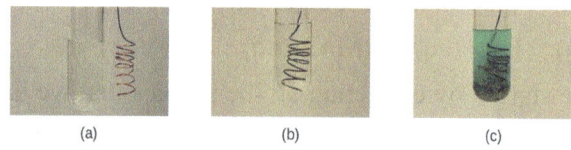

(a)                    (b)                    (c)

**Figure 6.3.** Galvanic reactions involving copper.

When a clean piece of copper metal is placed into a clear solution of silver nitrate (a), an oxidation-reduction reaction occurs that results in the exchange of Cu2+ for Ag+ ions in solution. As the reaction proceeds (b), the solution turns blue (c) because of the copper ions present, and silver metal is deposited on the copper strip as the silver ions are removed from solution. (credit: modification of work by Mark Ott)

Let us consider what happen when a clean piece of copper metal is added in a solution of silver nitrate. As soon as the copper metal is added into the solution, silver metal starts to form and also, the copper ions pass into the solution.

The presence of cooper ions in the solution is indicated by the blue colour of the solution on the far right. The reaction can be divided into its two half-reactions. Half-reactions separate the oxidation from the reduction, so each can be considered individually.

oxidation: Cu(s)→Cu2+(aq) +2e−

reduction:2×(Ag+(aq) +e−→Ag(s)) or 2Ag+(aq) +2e−→2Ag(s)

overall:2Ag+(aq) +Cu(s)→2Ag(s)+Cu2+(aq)

For the reduction half-reaction, the equation had to be doubled so the number electrons that are "gained" in the reduction half-reaction equals to the number of electrons that are "lost" in the oxidation half-reaction.

Spontaneous electrochemical reactions are involved in Galvanic or voltaic cells. In this, the half-reactions are separated so that current can easily flow all the way through an external wire. The beaker that is on the left side of the figure is known as a half-cell. 1 M solution of copper (II) nitrate [Cu (NO3)2] with a piece of copper metal that is partly immersed in the solution is contained in this half-cell.

The copper metal in the beaker is an electrode. The copper is undergoing the process of oxidation. Thus, in the reaction the copper electrode acts as the anode. The copper electrode that is the anode is connected to a voltmeter with a wire. And, the other terminal of the voltmeter is connected by a wire to a silver electrode. The silver electrode is the cathode in the reaction because it is undergoing the process of reduction.

The half-cell that is on the right side of the figure includes silver electrode in a 1 M solution of silver nitrate (AgNO3). No current flows at this point. No flow of current means there is no significant movement of electrons through the wire. This is because the circuit is open.

The circuit is closed by the use of a salt bridge. As the circuit is closed, flow of current is maintained with moving ions. The salt bridge includes a nonreactive, concentrated, electrolyte solution like the sodium nitrate (NaNO3) solution that is used in the example. As there is the flow of electrons from left to right through the wire and electrode, nitrate ions (anions) pass through the porous plug. This porous plug is on the left into the copper (II) nitrate solution.

This helps in keeping the beaker, that is on the left side, electrically neutral by neutralizing the charge on the copper (II) ions. These copper ions are produced in the solution as the copper metal goes through the process of oxidization. At this same point of time, it has been observed that the nitrate ions are moving to the left side. On the other hand, sodium ions (cations) move to the right side. They move through the porous plug, and into the silver nitrate solution that is on the right.

These added cations replaced the silver ions. These silver ions were removed from the solution as they were reduced to silver metal. This keeps the beaker on the right electrically neutral. The compartments would not remain electrically neutral and also no significant current would flow in the absence of the salt bridge. On the other hand, if the two compartments are in direct contact to each other, there is no need to have a salt bridge. The time at the circuit is completed, the reading on the voltmeter is +0.46V, this is called the cell potential. The cell potential is produced when the two different metals are linked and is a measure of the energy per unit charge available with the help of oxidation-reduction reaction. The volt is the derived SI units for the electrical potential:

Volt $=V =kg/m2A/s3 = JA/s =JC$

In the above given equation, where A is the current in amperes and C is the charge in coulombs. Note that the volts must be multiplied by thee charge in coulombs (C) in order to acquire the energy in joules (J).

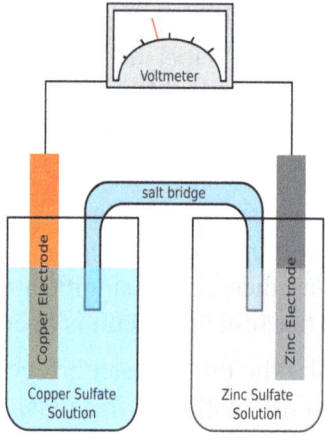

**Figure 6.4.** A typical Galvanic Cell.

In this standard galvanic cell, the half-cells are separated; electrons can flow through an external wire and become available to do electrical work.

When the electrochemical cells are constructed in this fashion, a positive cell potential shows an instant reaction and that the electron are moving from the left to the right direction. There is a lot happening in the figure mentioned above, in this way, it is very significant to summarize several things for this system:

- Electrons that are moving from the anode to the cathode: left to right in the standard galvanic cell as shown in the above-mentioned figure.

- The electrode that is present in the left half-cell is the anode, due to the oxidation takes place here. The name suggests to the flow of anions in the salt bridge in the direction towards it.

- The electrodes that are present in the right half-cell is cathode, due to reduction takes place here. And the name refers to the movement of the cations in the salt bridge towards it.

- The process of oxidation takes place at the anode (the left half-cell as shown in the figure).

- The process of reduction takes place at the cathode (the right half-cell as shown in the figure).

- The cell potential, +0.46V, in this case, as an outcome, from the inherent variations in the nature of the material used to make the two half-cells.

- The salt bridge must be present to close (complete) the circuit and both an oxidation as well as reduction must take place for current to move.

There are several numbers of possible galvanic cells, in this way, a shorthand notation is generally utilized in order to explain them. The cell notation (every so often known as cell diagram) provides information about the several numbers of species that took participation in the reaction. The notation also works for the other kinds of cells as well.

A vertical line, |, depicts a phase boundary and a double line, ||, depicts the salt bridge. Information about the anode is written to the left, which is followed by the anode solution, then the salt bridge (when present), then the cathode solution, and, in the end, information about the cathode to the right. The cell notation for the galvanic cell in Figure mentioned above can be written as

$Cu(s)|Cu2+(aq, 1M) ||Ag+(aq, 1M) |Ag(s)$

Note that spectator ions are not contained within and that the simplest form of each half-reaction was used. The initial concentrations of the several numbers of ions are generally included.

One of the simplest cells is the Daniell cell. It is possible to develop this kind of battery with the help of placing a copper electrode at the bottom of a jar and covering the metal with a copper sulfate solution. The solution of zinc sulfate is floated on the top of the solution of the copper sulfate, after then, a zinc electrode is placed in the solution of zinc sulfate.

Connecting the copper electrode to the zinc electrode permits the movement of the electric current. This is an example of a cell without the help of a salt bridge, and ions can flow across the interface between these two solutions.

## 6.4.3. The Nernst Equation

This topic will extend the study of electrochemistry by finding out the relationship between Excel and the thermodynamics quantities for example as $\Delta G°$ (that is Gibbs free energy) and K (which is the equilibrium constant). With respect to the galvanic cells, chemical energy is transformed into the electrical energy, which can be utilized to do work. The electrical work is the product of the charge transferred multiplied by the potential difference (that is voltage):

electrical work=volts × (charge in coulombs) = J

The charge on 1 mole of electrons is given by Faraday's constant (F)

$F = 6.022 \times 1023 e - mol \times 1.602 \times 10 - 19 Ce - =$

$9.648 \times 104 Cmol = 9.684 \times 104 JV \cdot mol$

total charge = (number of moles of e−) × F= nF

In the above-mentioned equation, n is the number of moles of electrons for the balanced reaction of oxidation and reduction. The measured cell potential is the maximum capability or energy that a cell can deliver and is connected to the electrical work (wele) by

Ecell= −wele/nF or wele= −nFEcell

The negative sign for the work, depicts that the electrical work can be done with the help of the system (which is the galvanic cell) on the surroundings. In the previous topic of the chapter, the free energy was defined as the energy which is available in order to do work. More specifically, the modification or alteration in the free energy was defined in the terms of the maximum work (wmax), which, for electrochemical systems, is denoted by the wele.

$\Delta G$= wmax = wele

$\Delta G$= −nFEcell

The signs can be verified whether they are correct or not, if n and F are positive constants and that galvanic cells, which have the positive cell potentials, which is consisting of the instant reactions. In this way, the spontaneous reactions, which have $\Delta G < 0$, must have Ecell> 0. If all the reactants as well as the products are in their standard states, in this way, the equation can be written as:

$\Delta G° = −nFE°cell$

This provides a way which is linked to the standard cell potentials to equilibrium constants, since

$\Delta G^\circ = -RT\ln K$

$-nFE^\circ cell = -RT\ln K$

Or $E^\circ cell = (RT/nF)\ln K$

Most of the time, the electrochemical reactions are run at standard temperature (298.15 K). Collecting terms at this temperature yields

Excel $= (RT/nF)\ln K = [\{(8.314\ JK/mol) * (298.15K)\}/n\times96{,}485*C/Vmol]\ \ln K$

$$= (0.0257V/n)\ln K$$

In the above equation, n is the number of moles of the electrons. From the observations made by the scientists in earlier times, the logarithm in the equations which is having the cell potential is every so often expressed with the help of base 10 logarithms (log), that alters the constant by a factor of 2.303:

Excel $= (0.0592\ V\ /n)\log K$

In this way, if $\Delta G^\circ$, K, or Excel is known or can be evaluated, the other two quantities can be readily determined. The relationships are shown graphically in Figure mentioned below.

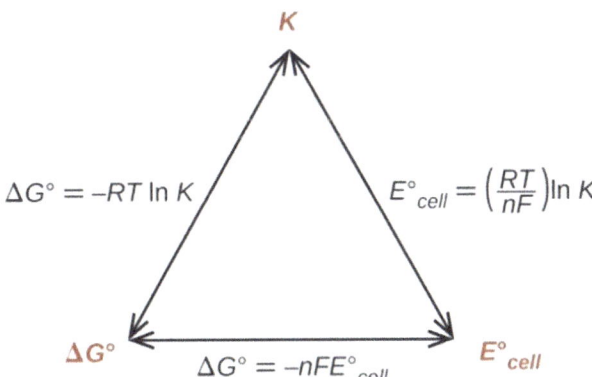

**Figure 6.5.** Relationship between Excel an Change in Gibbs Free Energy.

Given any one of the quantities, the other two can be evaluated.

## 6.5. BATTERIES AND FUEL CELLS

A battery is a series of cells or an electrochemical cell, which develops an electric current. Ideally, any kind of galvanic cell can be applied in order to create current from a battery. An ideal battery would develop an unchanging voltage, never run down, and having the capacity of enduring the extreme settings of humidity and heat.

Real battery provides a balance between the ideal aspects and the practical constraints. For instance, the weight of a normal car battery is near bout eighteen kilo grams or near about one per cent of the weight of a light-duty truck or an average car. This kind of battery can provide approximately unlimited energy if the battery is used in a smartphone, but it would be rejected for the application, and the reason behind that is the weight or mass of the battery.

In this way, no single battery is best. Batteries are selected for a specific type of application, it helps in keeping things such as the weight of the battery, reliability of the battery, cost of the battery, and capacity of current that it is producing. Basically, there are two kinds of batteries that are being used. First is primary battery and second id secondary battery. Some of the batteries of each types are described in the below mentioned topics.

## 6.5.1. Primary Batteries

The batteries that can be used only for a single time, are known as primary batteries. Primary batteries cannot be recharged. A typical example of primary battery is the dry cell (Figure given below). The dry cell is a zinc-carbon battery. The zinc can work either ways, as a container as well as the negative electrode.

The positive electrode, that is made of carbon which is surrounded with the help of a zinc chloride (Cl), paste of manganese (IV) oxide, carbon powder, and a small quantity of H2O. The reaction that take place at the anode can be depicted as the ordinary oxidation of zinc.

$Zn(s) \rightarrow Zn2+(aq) + 2e-$; $E°Zn2+/Zn=-0.7618$ V

The reaction that take place at the cathode is comparatively complex, in part because more than one reaction happens. The sequence of the reactions that take place at the cathode is near about:

$2MnO2(s)+2NH4Cl(aq)+2e- \rightarrow Mn2O3(s)+2NH3(aq)+H2O(l)+2Cl-$

The overall reaction for the zinc–carbon battery can be depicted as:

$2MnO2(s)+2NH4Cl(aq)+Zn(s) \rightarrow Zn2+(aq) +$
$Mn2O3(s)+2NH3(aq)+H2O(l)+2Cl-$

With an overall capability or potential of the cell, which is initially near about 1.5V, but as the cell put into use, its potential decreases. It is very necessary to remember that the voltage that is delivered by the battery in same regardless of the size of the battery. According to this reason, AAA, AA, A, D, and C batteries all have the similar rating with respect to the voltage.

Nevertheless, the large batteries can deliver more moles of the electrons. As the zinc container oxides, the contents of the same eventually lek out. In this way, this type of battery should not be left in any kind of electrical device for extended periods.

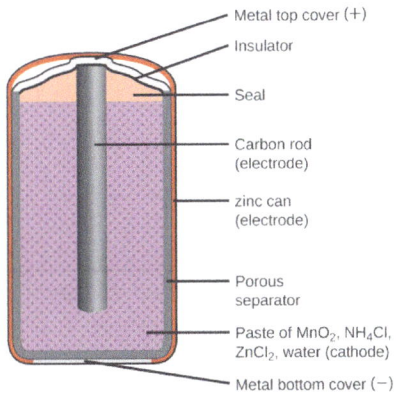

**Figure 6.6.** The diagram shows a cross section of a flashlight battery, a zinc-carbon dry cell.

*Source: Image by Wikimedia commons.*

Alkaline batteries (as illustrated in Figure 6.7) were developed in 1950s partly to cope up some of the performance issues with the zinc-carbon dry cells. They are manufactured to be the exact substitute with respect to the zinc-carbon dry cells. As their name recommends, these kinds of batteries utilize potassium hydroxide sometimes, but mainly they use alkaline electrolytes. The reaction involved are:

anode: $Zn(s)+2OH-(aq) \rightarrow ZnO(s)+H2O(l)+2e-$ E°anode=−1.28 V

cathode: $2MnO2(s)+H2O(l)+2e-\rightarrow Mn2O3(s)+2OH-(aq)$
E°cathode=+0.15 V

overall: $Zn(s)+2MnO2(s) \rightarrow ZnO(s)+Mn2O3(s)$    Excel=+1.43V

An alkaline battery can provide near about three to five times more the energy as compared to the zinc-carbon dry cell with same size. Alkaline batteries are likely to leak potassium hydroxide, in this way, these should also be removed from the devices with respect to the long-term storages.

On the other hand, there are few of the batteries that are rechargeable, and also, most are not. Attempts to recharge an alkaline battery that is not rechargeable every so often leads to tear of the battery and the leakages of the potassium hydroxide electrolyte.

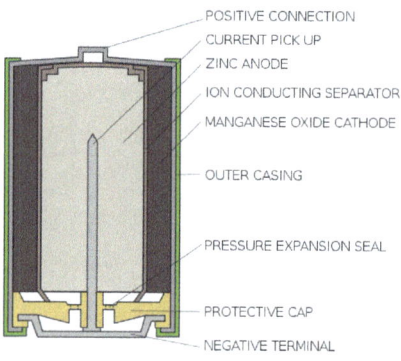

POSITIVE CONNECTION
CURRENT PICK UP
ZINC ANODE
ION CONDUCTING SEPARATOR
MANGANESE OXIDE CATHODE

OUTER CASING

PRESSURE EXPANSION SEAL

PROTECTIVE CAP

NEGATIVE TERMINAL

**Figure 6.7.** Alkaline batteries were designed as direct replacements for zinc-carbon (dry cell) batteries.

*Source: Image by Wikimedia Commons.*

## 6.5.2. Secondary Batteries

Secondary batteries are rechargeable in nature. the secondary batteries are the kinds of batteries that are found in devices such as electronic tablets, smartphones, and automobiles.

Nickel-cadmium batteries or NiCd batteries (as shown in the Figure 6.8) is consisting of a which is nickel-plated, anode is plated with cadmium, and a potassium hydroxide electrode. The positive plate and the negative plates, that are prevented from shorting with the help of a separator, are rolled together and put into the case. This is a "jelly-roll" design and permits the NiCd cell to provide much more current as compared to the alkaline battery of same size. The reaction involves:

anode:  $Cd(s)+2OH^-(aq) \rightarrow Cd(OH)2(s)+2e-$

cathode:  $NiO2(s)+2H2O(l)+2e-\rightarrow Ni(OH)2(s)+2OH^-(aq)$

overall:  $Cd(s)+NiO2(s)+2H2O(l) \rightarrow Cd(OH)2(s)+Ni(OH)2(s)$

The amount of voltage that these batteries produces is near about 1.2V and 1.25V. When secondary batteries or Nickel-cadmium batteries are treated in a proper manner, a NiCd battery can be recharged near about a thousand times. Cadmium is a toxic heavy metal, in this way, the Nickel-cadmium batteries should never be opened or put into the regular trash.

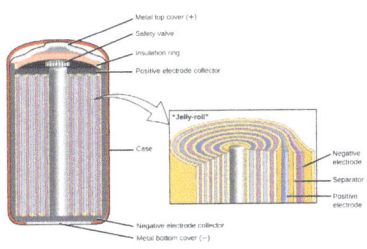

**Figure 6.8.** NiCd batteries use a "jelly-roll" design that significantly increases the amount of current the battery can deliver as compared to a similar-sized alkaline battery.

*Source: Image by Wikimedia Commons.*

Lithium ion batteries: (as shown in the Figure 6.9) are among the most common rechargeable batteries and these batteries are used in several numbers of portable electronic devices. The reactions involved are:

anode:        $LiCoO2 \rightleftharpoons Lix-1CoO2+xLi^{+}+xe^{-}$

cathode:      $xLi^{+}+xe^{-}+xC6 \rightleftharpoons xLiC6$

overall:      $LiCoO2+xC6 \rightleftharpoons Lix-1CoO2+xLiC6$

4With the coefficient representing moles, x is no more than near about 0.5 moles. The battery voltage is about 3.7V. Lithium batteries are most common and the reason behind this is they can deliver huge amount of current are lighter as compared to the batteries of other types, they approximately produce a constant voltage as they discharge, and only slowly lose their charge when stored.

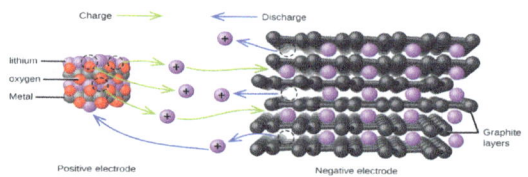

**Figure 6.9.** In a lithium ion battery, charge flows between the electrodes as the lithium ions move between the anode and cathode.

*Source: Image by Wikimedia Commons.*

### 6.5.3. Fuel Cells

A fuel cell is a device that transforms the chemical energy into the electrical energy. Fuel cells are very identical to the batteries, but they necessitate a continuous source of fuel, sometimes it is hydrogen. They will continue to develop the electricity as long as fuel is available. Cells made of hydrogen fuel have been used in order to supply the power for the space capsules, boats, satellites, automobiles, and submarines (as shown in the Figure 6.7).

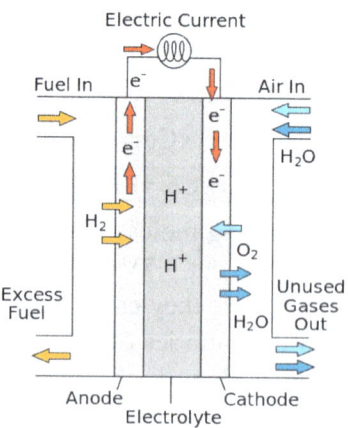

**Figure 6.10.** In this hydrogen fuel-cell schematic, oxygen from the air reacts with hydrogen, producing water and electricity.

*Source: Image by Wikimedia Commons.*

Today, fuel cells are regarded as one of the promising sources for renewable energy because of its ability to produce environment friendly and recoverable by products. This is because the efficiency of fuel cells can be extended up to higher than 50% while internal combustion engines today ate only at around 38-40%. In addition, fuels cell's engineering does not have moving parts that reduces further its efficiency. Similarly, fuel cells rely on reversible electrochemical reactions that can be optimized at lower or even higher temperatures of 1000 K or higher.

In a hydrogen fuel cell, the reactions are

anode:           $2H_2 + 2O_2- \longrightarrow 2H_2O + 4e-$

cathode:         $O_2 + 4e- \longrightarrow 2O_2-$

overall:         $2H_2 + O_2 \longrightarrow 2H_2O$

The voltage is produced is near about 0.9V. the effectiveness of fuel cells is generally about 40–60%, which is higher as compared to the conventional internal combustion engine (25–35%).

On the other hand, in the case of the hydrogen fuel cell, they produce only water as exhaust. In the present, fuel cells are rather expensive and is consisting of features that cause them to fail after a comparatively short time.

Three main mechanisms were cited to be of importance in hydrogen fuel cells. The hydrogen evolution reaction (HER) which is facilitated by the cathode, the oxygen evolution reaction (OER) facilitated by the anode and the ion activity and mobility as facilitated by the working electrolyte. These reactions are usually the ones being considered in the design of fuel cells.

One of the promising advancements in the field fuel cells are solid oxide fuel cells (SOECs). Solid oxide fuel cells can be simply described as the reversible process of fuel cells that utilizes sold oxides as electrodes (Solid Oxide Fuel Cells – SOFCs). With SOECs the anode of the fuel cell becomes its cathode and the cathode of the conventional fuel cells becomes its anode – interchanging the HER and OER. Today, the use of ceramic metals oxides for SOECs make it possible to build a fuel cell made up of solid oxides at ceramic metals in a pure solid-state form. This engineering of SOECs allow it operate at high temperatures stating 500K to almost above 100K. Typical fuel cells operate until about 150 degrees C, while SOFC can extend to 450°C (Mendoza, 2017).

# 6.6. APPLICATIONS OF ELECTROCHEMISTRY H2O

There are several different uses of electrochemistry, particularly in industry. The principles of cells are used in order to make electrical batteries. In the context of technology, a battery is defined as a device that stores chemical energy and makes this energy available in an electrical form. Electrochemical devices such as one or more galvanic cells or fuel cells are used to make batteries. There are many small scales uses of including in:

- Electrical appliances like cell phones (long-life alkaline batteries)
- electrical Torches
- Hearing aids (silver-oxide batteries are used)
- Digital watches (mercury/silver-oxide batteries are used)
- Military applications (thermal batteries are used)
- Digital cameras (lithium batteries)

Large scale applications of batteries are usually for off-grid energy storage and emergency power generation, hybrid and electrical vehicles.

## 6.6.1. Electroplating

The process of electroplating takes place when an electrically conductive object is covered with a layer of metal using electrical current. The electrolytic cell can be used for the process of electroplating.

It has been observed that at Times, electroplating is used in order to give a metal particular property or it is also used for aesthetic reasons:

- The production of jewellery
- Abrasion and wear resistance-
- Corrosion protection-

**Figure 6.11.** An illustration of electroplated piece of aluminum artwork.

*Source: Image by Flickr.*

Electro-refining is also known as electrowinning is electroplating on a large scale. Copper plays a very important role in the electrical industry. This is because it is very conductive and is also used in electric cables. One of the issues with cooper is that it must be pure if it is to be an efficient current carrier.

Electrowinning is one of the methods that is used to purify copper. In this method, copper ore is processed into impure blister copper. Then, it is deposited as pure copper through the process of electroplating. The process of copper electrowinning is defined below:

- There is a bar of impure copper that contains other metallic impurities. This acts as the anode.
- The cathode is made up of pure copper. It also has few impurities.
- The electrolyte is a solution of aqueous $CuSO_4$ and $H_2SO_4$.
- Electrolysis takes place when the current passes through the cell:
  - The impure copper anode is oxidized to form $Cu2+$ ions in solution.
  - The anode decreases in mass. $Cu(s) \rightarrow Cu2+(aq) + 2e-$
  - At the cathode, reduction of positive copper ions occurs to make pure copper metal.
  - The cathode increases in mass. $Cu2+(aq) + 2e- \rightarrow Cu(s)$ (>99% purity)
- The other metal impurities that are present in the solution do not dissolve (Au(s), Ag(s)). There is the formation of a solid sludge at the bottom of the tank or remain in solution (Zn(aq), Fe(aq) and Pb(aq)) in the electrolyte.

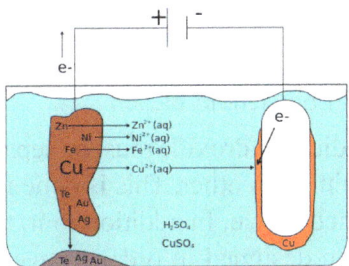

**Figure 6.12.** A simplified diagram to illustrate what happens during the electro-winning of copper.

*Source: Image by Wikimedia Commons.*

## 6.6.2. The Chloralkali Industry

The chlorine-alkali (chloralkali) industry is a very important part of the chemical industry. It produces chlorine as well as sodium hydG7roxide through the electrolysis of the brine. Brine is a raw material. It is saturated solution of sodium chloride (NaCl) and is obtained from natural salt deposits.

There are numbers of important uses of the products of the chloralkali industry:

It is used:

- as a disinfectant
- For H2O purification
- In the production of:
    - Paints, solvents, plastics (such as polyvinyl chloride), petroleum products;
    - hypochlorous acid (used to kill bacteria in drinking H2O)
    - antiseptics, textiles, insecticides, laboratory chemicals, medicines
    - paper and food

Sodium hydroxide is also known as caustic soda. It is used:

- To make rayon (that is artificial silk)
- For the purification bauxite (the ore of aluminium)
- For the preparation of soap and other cleaning agents
- To make paper

One of the problems while producing chlorine as well as sodium hydroxide is that when they are produced together, the chlorine gets mixed up with the sodium hydroxide to form chlorate (ClO− and chloride (Cl−) ions. As a result of this, sodium chlorate, NaClO, is produced. It is one of the major components of household bleach.

The chlorine and sodium hydroxide must be separated from each other so that they do not react with each other. This is done in order to overcome the problem that is mentioned above. In addition to it, there are three industrial processes that have been designed in order to overcome this problem. All three methods include electrolytic cells.

The Mercury Cell is shown in the figure mentioned below:

- Carbon electrode is the anode. It is suspended from the top of a chamber.

- Liquid mercury is the cathode. It flows along the floor of this chamber.

- The electrolyte is brine (that is NaCl solution). This solution is passed through the chamber.

- Chloride ions in the electrolyte are oxidized at the anode when an electric current is applied to the circuit. Chloride ions are oxidized to form chlorine gas.

$2Cl^-(aq) \rightarrow Cl2(g) + 2e^-$

- At the same time sodium ions are reduced to solid sodium at the anode. The solid sodium gets dissolves in the mercury. Thus, making a sodium/mercury amalgam.

$Na^+(aq) + Hg(l) + e^- \rightarrow Na\ (Hg)$

- The amalgam is poured into a separate vessel. In this vessel, it decomposes into sodium and mercury.

- Inside the vessel, the sodium reacts with H2O and it also produces sodium hydroxide and hydrogen gas. On the other hand, the mercury returns to the electrolytic cell to be used again.

$2Na\ (Hg) + 2H2O(l) \rightarrow 2NaOH(aq) + H2(g) + Hg(l)$

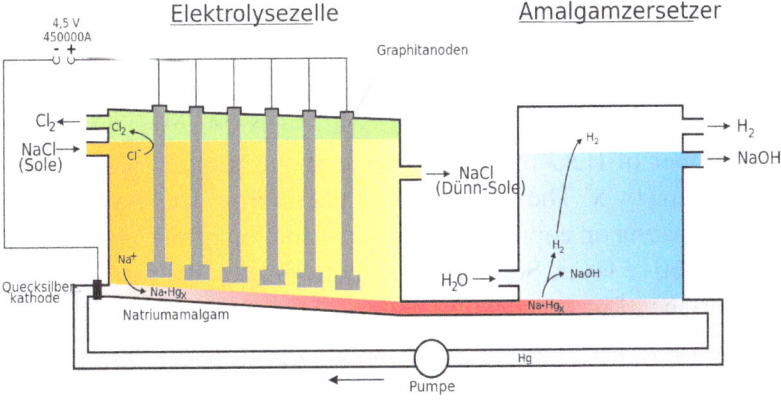

**Figure 6.13.** The mercury cells.

*Source: Image by Wikimedia Commons.*

## 6.7. CORROSION

Corrosion is generally defined as the degradation of metals. This degradation is because of an electrochemical process. There are various examples of corrosion. Some of them include tarnish on silver, the formation of rust on iron and the blue-green paint that develops on copper.

In the United States, the total cost of corrosion is significant, with estimations in excess of half a trillion dollars a year.

The formation of rust on iron is perhaps the most familiar example of corrosion. Rust is formed on iron when it is exposed to oxygen and $H_2O$. The main steps that are involved in the rusting of iron are as follows (Figure 6.14). Iron rapidly oxidizes once it is exposed to the atmosphere.

anode: $Fe(s) \longrightarrow Fe2+(aq) + 2e-$
$E°Fe2+/Fe = -0.44$ V

Oxygen is reduced by the electrodes in the air in acidic solutions.

cathode: $O2(g) + 2H+(aq) + 4e- \longrightarrow 2H2O(l)$
$E°O2/O2 = +1.23$ V

overall: $2Fe(s) + O2(g) + 4H+(aq) \longrightarrow 2Fe2+(aq) + 2H2O(l)$
Excel $= +1.67$ V

In scientific terms, rust is known as hydrated iron (III) oxide. This is formed forms when iron (II) ions react further with oxygen.

$4Fe2+(aq) + O2(g) + (4 + 2x) H2O (l) \longrightarrow 2Fe2O3·xH2O(s) + 8H+(aq)$

The number of $H_2O$ molecules in the above reaction is variable, so it is denoted there by x. The formation of rust on iron does not resemble the formation of patina on copper. In fact, the formation of rust on iron does not create a protective layer. So, iron keeps corroding as the rust gets removed from the surface of iron and exposes fresh iron beneath to the atmosphere.

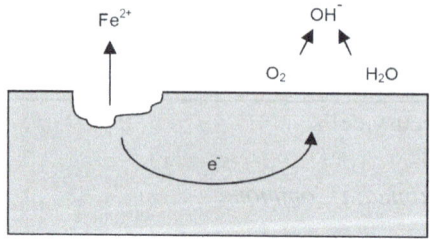

**Figure 6.14.** Representation of corrosion of metals.

*Source: Image by Wikimedia.*

Once the paint is scratched on a painted iron surface, corrosion occurs, and rust begins to form. The speed of the spontaneous reaction is increased in the presence of electrolytes, such as the NaCl used on roads to melt ice and snow or in salt H2O.

One of the methods that can be used to keep iron from corroding is to keep the iron painted. The layer of paint on the iron successfully prevents the H2O and oxygen to come in contact with the iron. Thus, the iron gets protected from corrosion as H2O and oxygen must come in contact with iron to corrode it. The iron is protected from corrosion as long as the paint remain intact.

There are several other strategies that can be used to overcome this problem. The other strategies involve alloying the iron with other metals. For instance, stainless steel is mainly iron with a bit of chromium. In stainless steel, the chromium tends to collect near the surface. Here, it forms an oxide layer. This layer helps in protecting the iron.

A different strategy is used in case of zinc-plated or galvanized iron. Zinc is more easily oxidized as compared to that of iron. The reason is because zinc has a lower reduction potential. Zinc is a more active metal since it has a lower reduction potential. Therefore, even if the coating on the zinc is scratched, the zinc will still oxidize before the iron oxidize. This shows that this strategy should work with other active metals.

There are several other ways to protect metal. One of the important ways to protect metal is to make it the cathode in a galvanic cell. This process of protecting the metal is known as cathodic protection. It can be also be used for metals other than just iron. For instance, the formation of rust on the underground iron storage tanks and pipes can be prevented or it can be greatly reduced by connecting them to a more active metal like zinc or magnesium.

Cathodic protection is used in many other metals. It is also used to protect the metal parts that are present in H2O heaters. The more active metals those who have lower reduction potential are known as sacrificial anodes because as they get used up as they corrode (oxidize) at the anode.

The metal that is being protected acts as the cathode, and so it does not oxidize (that is corrode). When the anodes are monitored properly and it is also periodically replaced, the useful lifetime of the iron storage tank can be greatly extended.

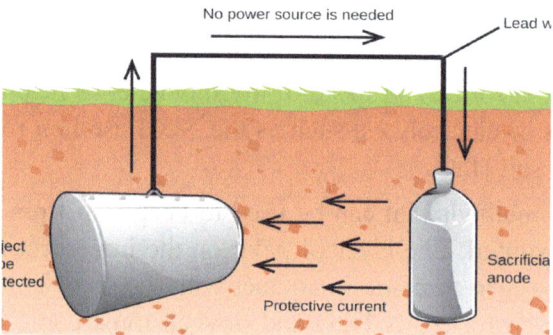

**Figure 6.15.** An underground iron storage tank is used in cathodic protection.

*Source: Image by Wikimedia Commons.*

One way to protect an underground iron storage tank is through cathodic protection. Using an active metal like zinc or magnesium for the anode effectively makes the storage tank the cathode, preventing it from corroding (oxidizing)

## 6.8. CONCLUSION

Material processing has been the most important industry for the survival of human beings. Every physical article or product that is being used by the human beings is directly or indirectly related to the material processing industry. This should be noted that all the procedures in material processing industry are based on maintaining the equilibrium. Various phase diagrams are used to explain the process of maintain the equilibrium, especially balancing the vapor-liquid equilibrium. There are a number of applications of electrochemistry that are very useful in the modern world. This should be noted that with the help of batteries and fuel cells, the material processing takes place.

There are two major types of batteries that are used in the physical world. It should be noted that these batteries and fuel cells also derive their energy from combination of different chemicals and compounds. There is one important phenomenon that explains the material processing industry. This phenomenon is corrosion. Corrosion explains how an object can be protected from being damaged by the practical application of material processing procedures and how thermochemistry plays an important role in carrying out the procedures of material processing industry.

# REFERENCES

1. Francis, L., (2016). Introduction to Materials Processing. Materials Processing, [online] pp.1–20. Available at: https://www.sciencedirect.com/science/article/pii/B978012385132100001X (accessed on 13 March 2020).

2. Lumen learning, (n.d.). Balancing Oxidation-Reduction Reactions | Chemistry for Majors. [online] Courses.lumenlearning.com. Available at: https://courses.lumenlearning.com/chemistryformajors/chapter/balancing-oxidation-reduction-reactions/ (accessed on 13 March 2020).

3. Lumen learning, (n.d.). Batteries and Fuel Cells | Chemistry for Majors. [online] Courses.lumenlearning.com. Available at: https://courses.lumenlearning.com/chemistryformajors/chapter/batteries-and-fuel-cells-2/ (accessed on 13 March 2020).

4. Lumen learning, (n.d.). Corrosion | Chemistry for Majors. [online] Courses.lumenlearning.com. Available at: https://courses.lumenlearning.com/chemistryformajors/chapter/corrosion/ (accessed on 13 March 2020).

5. Lumen learning, (n.d.). Galvanic Cells | Chemistry for Majors. [online] Courses.lumenlearning.com. Available at: https://courses.lumenlearning.com/chemistryformajors/chapter/galvanic-cells/ (accessed on 13 March 2020).

6. Lumen learning, (n.d.). Introduction to Electrochemistry | Chemistry for Majors. [online] Courses.lumenlearning.com. Available at: https://courses.lumenlearning.com/chemistryformajors/chapter/introduction-to-electrochemistry/ (accessed on 13 March 2020).

7. Lumen learning, (n.d.). Introduction to Thermochemistry | Chemistry for Majors. [online] Courses.lumenlearning.com. Available at: https://courses.lumenlearning.com/chemistryformajors/chapter/introduction-to-thermochemistry/ (accessed on 13 March 2020).

8. Lumen learning, (n.d.). The Nernst Equation | Chemistry for Majors. [online] Courses.lumenlearning.com. Available at: https://courses.lumenlearning.com/chemistryformajors/chapter/the-nernst-equation/ (accessed on 13 March 2020).

9. Mendoza, RMO, Cervera, RBM, Chuang, P-YA. A current-voltage model for hydrogen production by electrolysis of steam using Solid Oxide Electrolysis Cell (SOEC). 15th International Conference on

Environmental Science and Technology, September 2017.

10. Siyavula, (2020). Applications of Electrochemistry | Electrochemical Reactions | Siyavula. [online] Siyavula.com. Available at: https://www.siyavula.com/read/science/grade-12/electrochemical-reactions/13-electrochemical-reactions-07 (accessed on 13 March 2020).

11. Wattco, (2020). Phase Equilibrium – Wattco. [online] Wattco. Available at: https://www.wattco.com/casestudy/phase-equilibrium/ (accessed on 13 March 2020).

12. Lumen learning, (n.d.). Energy Basics | Chemistry for Majors. [online] Courses.lumenlearning.com. Available at: https://courses.lumenlearning.com/chemistryformajors/chapter/energy-basics/ (accessed on 13 March 2020).

13. Plascencia, G., & Jaramillo, D., (2017). Basic Thermochemistry in Materials Processing. [online] Available at: https://link.springer.com/book/10.1007/978–3-319–53815–0#about (accessed on 13 March 2020).

14. Wps.prenhall.com. (2010). Thermochemistry. [online] Available at: http://wps.prenhall.com/esm_hillpetrucci_genchem_4/0,8603,1079637-,00.html (accessed on 13 March 2020).

# Computational Thermochemistry and Its Relation to Thermodynamics

## CONTENTS

Computational thermochemistry is a subject that involves the use of a computer to make thermochemical predictions. The computational thermochemistry makes use of empirical methods that require only an instant on a slow personal computer. The chapter focuses on the meaning and methods for making predictions of the computational thermochemistry. In the mid half of the chapter, the challenges and outlook of the computational thermochemistry is discussed. The relation and the difference between thermodynamics and computational thermochemistry has also been given importance in the end of the chapter.

## 7.1. THERMOCHEMISTRY

Lavoisier and Laplace (1784) realized the need and importance of thermal phenomena in chemical reactions. The chemists laid the foundation of thermochemistry to address these needs. The study of affinity was done from the aspect through the attempts to find a measure of the chemical forces by the amount of heat given out in a chemical reaction.It was assumed that the amount of heat evolved in a chemical reaction is equal to the heat absorbed in the reverse reaction. In the combustion and respiration, measurement was done in some specific heats and the amount of heat evolved in reactions,Lavoisier's 'caloric' theory of combustion and Berzelius's electrochemical theory was considered as unsatisfactory by Persoz. He also emphasized that there is no means of explaining the heat developed in chemical reactions.The study of the heat released or absorbed as a result of chemical reactions is called thermochemistry. Thermochemistry is widely in use by the range of scientists and engineers. It is a branch of thermodynamics.

Figure 7.1. Chemical reaction is accompanied by release or absorption of heat.

*Source: Image by tcu.*

The applications of thermochemistry are:

1.   Thermochemistry is used by the biochemists to understand bioenergetics.

2.   Thermochemistry is applied by the chemical engineers while designing manufacturing and industrial plants.

In chemical reactions, there is a conversion of a set of substances referred to as "reactants" to a set of substances referred to as "products." The following balanced chemical reaction shows that the reactants are gaseous methane, $CH_4(g)$, and gaseous molecular oxygen, $O_2(g)$, and the products are gaseous carbon dioxide, $CO_2(g)$, and liquid water $H_2O(l)$:

$$CH_4(g) + 2O_2(g) \rightarrow CO_2(g) + 2H_2O(l) \quad (1)$$

There is change every time a chemical reaction takes place. Sometimes the change occurs on its own. The process of change that occurs on its own is called the spontaneous. However, the change that does not occurs on its own is called non-spontaneous.

For example, if a barrel of fuel is left alone, it will be fuel only. However, if match is used to ignite the fuel, the fuel will burn spontaneously till the time all the reactants air, fuel is completely consumed.

Under this instance, there was requirement of a small amount of energy to be added to the system before a much larger amount of energy could be released in the spontaneous process. However, once the spontaneous process starts, it proceeds without any assistance. In an electrolysis reaction, where the electricity is passed through $H_2O$ to dissociate it into hydrogen and oxygen, it is not considered spontaneous because the reaction stops taking place if the electricity is removed.

# 7.2. COMPUTATIONAL THERMOCHEMISTRY

Any method that involves the use of a computer to make thermochemical predictions is a computational thermochemistry.

The computational thermochemistry makes use of empirical methods that require only an instant on a slow personal computer. Also, high-level quantum calculations require the most powerful supercomputers.

This chapter emphasizes more on gas-phase enthalpies of formation. However, most practical work is not in the gas phase but in the condensed phase, such as in solution and at surfaces. There are instances, gas-phase values being adequate, the thermochemistry of a chemical reaction is not very sensitive to the solvent.

In today's world, computer-assisted thermochemistry can be applied in many fields today. Particularly, with the use of reliable thermochemical source data and appropriate application software, reacting amounts and gas pressures, optimum operating temperatures necessary to get the product of the required purity can be obtained. Therefore, the time-consuming experimental work and the cost of that work can be reduced.

The calculation of thermochemical equilibrium states forms the basis for the solution of many problems in process engineering or materials science. The Gibbs energy of the system must be minimized while calculations.

In the chemical sciences, knowledge of the thermochemistry of molecules has major importance and it is essential to many technologies. Information on stabilities and reactivities of molecules that are used can be found through the thermochemical data. For example, In combustion, in modelling reactions, the atmosphere, and chemical vapor deposition.

In the chemical industry, for the safe and successful scale up of chemical processes, thermochemical data plays an important role. There are numerous species for which there are no data, in spite of the fact that there was compilation of experimental thermochemical data for many molecules.

However, it has been found that the data in the compilations are sometimes incorrect. Also, the experimental measurements of thermochemical processes are expensive and difficult, that is why computational methods are highly desirable for making reliable predictions.

One of the major goals of modern quantum chemistry is the prediction of molecular thermochemical data to chemical accuracy ($\pm 1$ kcal/mol), in the early 1970s, when ab-initio molecular orbital (MO) calculations became routine.

In the 1970, the problem aroused in the Hartree-Fock (HF) calculations which gave large errors of up to 100 kcal/mol in bond energies. In order to correctly predict the thermochemical data, it is required to go beyond the Hartree-Fock theory for including a sophisticated treatment of electron correlation, which made the calculations very difficult.

With the advancement in the theoretical methodology, development of computer algorithms, and increases in computer power, there has been considerable progress made in attaining the goal of a $\pm 1$ kcal/mol. It has now made possible to calculate the reliable thermochemical properties for a fairly wide variety of molecules.

The methods that are currently being used for computing thermochemical data ranges from very high levels of theory to that of moderate levels of theory with some form of empirical input, at the ab initio MO level. The very high level of theory is limited to smaller molecules and can attain accuracies of ±0.5 kcal/ mol, while the latter methods can be applied to larger molecules with somewhat less accuracy.

There has been a widely used method for predicting the thermochemical data of molecules and the method is known as Gaussian-n methods. Later in this chapter, the current state-of-the-art methods in computational thermochemistry will be explained. The Gaussian-n approach and Gaussian-3 (G3) theory, which have achieved a new level of accuracy in the computational theory.

In Computational thermochemistry, there are various methods for computing thermochemical data that range from empirical schemes to ab initio MO theory. Ab Initio methods can be roughly divided into three types:

(1)   Very high-level quantum chemical calculations with no experimental input;

(2)   Composite techniques that combine moderate level ab initio quantum chemical calculations with some form of molecule-independent empirical parameters; and

(3)   Techniques that use molecule-dependent empirical parameters obtained from accurate experimental data in combination with moderate level ab initio quantum chemical calculations.

The purpose of computational thermochemistry is to summarize all the available methods for predictive thermochemistry.

Comparison of Methods

Table 7.1. Table Showing the comparison of methods

| Method | Applicability | |
| --- | --- | --- |
| | Type | Size |
| Empirical (e.g., Benson groups) | common organic | Any |
| Molecular mechanics (e.g., MM3) | common organic | 100,00 atoms |
| Semiempirical M O theory (e.g., MNDO/d) | organic, some inorganic | 500 atoms |

| Density functional theory (e.g., B3LYP) | All | 50 atoms |
|---|---|---|
| CBS-4 (M O theory) | All | 20 atoms |
| BAC-MP 4 (corr. M O theory) | organic, some inorganic | 20 atoms |
| PCI-80 (corr. M O theory) | All | 20 atoms |
| G2 (MO theory) | All | 6 heavies |
| CCSD(T) with basis set extrapolation (M O theory) | All | 3 heavies |

*Source: Table by pubs.acs.org.*

The above table represents different methods of computational thermochemistry. The selection of several predictive methods is compared in the above table on the basis of applicability. Applicability implies the type and size of molecule that a method can accommodate. This depends on the limit on molecule size that may be higher for symmetrical molecules. Also, the rankings in the "accuracy" and "reliability" depend upon the type of molecule for which the predictions are needed.

All the methods discussed above complements each other and all are useful, for example, the semi-quantitative comparisons given in the above table reflect the subjective judgments of people which is considered as "typical" applications.

## 7.2.1. Empirical Estimation Methods

When there is absence of molecule of interest in the experimental databases such as that by Pedley, empirical estimation scheme is used there. Empirical estimation scheme is very cheap to use. When there are necessary parameters are available, the more sophisticated schemes are generally very reliable.

Under this method, quality of the parameter values defines the quality of the results. When the parameter value is derived from the measurement of several different compounds, it is considered to be more certain than the value derived from a single measurement.

Bond additivity is the simplest and most useful method for predicting reaction enthalpies. This is based on the certain assumption that the crude approximation of all bonds of same type are equally strong.

Although, rough results can only be expected from this simple method. Bond additivity can be used in a form that leads to enthalpies of formation.

## 7.2.1.1. Group Additivity

Using the group additivity method, there are better results that can be obtained based on the functional group. There are two simple examples that are the methods due to Joback and to Cardozo. In order to use such method, it is required to identify and count the appropriate groups within the molecule. After that, sum their respective contributions. This is considered as an easy exercise and requires no nomenclature or special notation.

Sophisticatedly, the second-order group methods include the effects of the chemical environment (second-nearest neighbours) on a group's contribution. The second order group methods are the complicated and in terms of accuracy, it is better than the first order group method. However, the method requires some effort to learn. Most popular method is developed by Benson and co-workers.One of the major practical advantages of the Benson's method is that it has been incorporated into the computer programs. Therefore, it helps the user from doing the arithmetic or even identifying the groups and saves their time. There are various corrections that account for interactions such as steric crowding and ring strain.There are recent developments that has taken place and are described in this chapter by Benson and Cohen. Other than the Benson's approach, there is a second-order group methods that exist. Cox and Pilcher, in their classic book, pointed out the mathematical equivalency in the methods developed by Benson, by Laidler and by Allen, despite the fact that they employ different chemical interpretations of the interactions.

The value of the parameter and worked examples for these three methods can be found in Cox and Pilcher's book. In case where the calculation is done by hand, Reid et al. found the method by Yoneda to be comparable to Benson's in accuracy. It is to be noted that it did not require the calculation of symmetry numbers for entropy estimations.

Recently, Pedley developed a method for evaluation of experimental data in the literature is called estimation method. Sometimes, not all the group parameters are available for the molecule of interest. The most expedient solution is often to estimate the value for the missing group by comparison with similar groups.

## 7.2.1.2. Limitations of Group Additivity

As given in the table, group additivity schemes are restricted to common organic molecules. There are many other types of molecules present. As the case of the thermochemistry of gas-phase ions, it is not well-estimated using group additivity schemes.

Holmes and Lossing has developed a scheme for estimating cation thermochemistry and these are summarized as GIANT (Gas-Phase Ion and Neutral Thermochemistry) tables. Generally, these methods are based upon trends in ionization energies or proton affinities within homologous series of molecules.

Because of the absence of the group values of many inorganic and organometallic compounds, and especially transition-metal compounds, the estimation by standard group-based methods is not possible. More often, the thermochemistry of related organic compounds can be used to estimate enthalpies of formation of organometallics as few estimation schemes are available.

Electrostatic-covalent model, by Drago and Cundari which is described in the book is applicable to wide variety of compounds. These have been parametrized for thermochemical quantities such as the energetics of gas-phase ion-molecule reactions, the strengths of dative and covalent bonds, solvation energies.

Organometallic thermochemistry has a fundamental problem of frequently occurring difficulty of experimental measurements. Therefore, there are possibility of large errors reported in the values. Unfortunately, there is absence of the robust and sufficient method for estimating the thermochemistry of organometallic compounds.

## 7.2.2. Molecular Mechanics

A mechanical assembly of balls and springs (atoms and chemical bonds) with a few other forces added to simulate electrostatics, dispersion, is called molecule. Such molecules have widespread use in biological and pharmaceutical chemistry and are collectively described as molecular mechanics (MM).

Among the organic MM methods, only those are parametrized for enthalpies of formation which are from the Allinger group (e.g., MM3 and MM4). There is a need of computer in MM method, because all the geometric parameters of the molecule must be varied to minimize the molecule's computed energy. However, for calculations on organic molecules of typical size fewer than 500 atoms, there is need of only a desktop, personal computer.

Chemical Reasoning

The above systematic methods cannot be used in many cases, as some or the other parameters are lacking in every method explained above.

Therefore, there was need to construct one effective strategy in which a hypothetical reaction was made that satisfies two conditions:

(1)    It is approximately thermoneutral (i.e., A// 0 = 0),

(2)    thermochemical data are available for all the compounds involved except the one of interest

In order to meet the first requirement, there is need to have a correctness of judgment for the quality prediction. It should be noted that the chemical reasoning is useful for guessing the values of missing parameters.

## 7.2.3. Quantum Mechanical Models

Quantum Mechanical Method is used in the case when there is a lack of experimental data and even the empirical estimation methods fail. Free radicals or transition states are most often essential for calculating reaction rates.

The techniques discussed above model a molecule as an assembly of electrons and atomic nuclei. Generally, this is a good feature. However, computationally the quantum calculations are much more intensive than the empirical models. The experimental results can be exploited to correct a theoretical prediction. However, there has been a compromise between the reliability of the resulting prediction and the cost of the calculations.

Those empirical corrections should be incorporated within the semiempirical methods theory or can be applied to the final result.

Quantum chemical model are classified into three common types:

1.    Semiempirical molecular orbital (MO) theory;

2.    Ab initio (non-empirical) MO theory; and

3.    Density functional theory (DFT).

These three methods aim to solve the non-relativistic, electronic Schrodinger equation by minimizing the total energy of the molecule using various approximations. The lower-energy orbitals contain electrons and the higher-energy orbitals are empty. The concept of MO theory is based on the concept of orbitals.

DFT is based upon a theorem that the electronic energy depends only upon the electron density distribution within the molecule. Practically, in order to compute the electron density, DFT calculations often use MOs. DFT is conceptualized based on the early attempts to explain electronic structures of substances. Like how Erwin Schrodinger in 1925 have obtained the detailed electronic structure of benzene stating the "al the information is contained in the wave function of 120 coordinates and 42 electron spin components of benzene". This was followed on by the Born-Oppenheimer approximation in 1927, postulating that "most of the information we want to know about chemistry is in the electron density and electronic energy". The birth of DFT was probably in 1964 when Pierre, Hohenberg and Walter Kohn stated that "all the information is contained in density – a simple function of 3 coordinates". So far, to date, DFT is one of the most utilized approximations in the atomic level.

Five years ago, the field of quantum chemistry mostly used the MO theory and was fully dominated by the MO theory. Nowadays, DFT calculations are very popular for chemical applications.

## 7.2.4. Semi-empirical MO Theory

There are many interactions takes place as there are small molecules that contains many electrons.

Decades ago, the computers were not sufficient for the heavy calculations. They were much weaker and also the ab initio calculations were impossible for molecules containing more than a few atoms.Semi-empirical methods were developed where most of the interaction terms was neglected and replaced by the empirical parameters that resulted in reduction of the computational cost.For reproducing experimental results, there was some additions to the empirical forces for its aid. There are some other features that ab initio method also has, such as the use of an atomic-orbital basis set and the possibility of including electron correlation explicitly.

Semi-empirical methods have been ruling over many years. Nowadays, computers are very powerful and because of which the more rigorous ab initio methods have become practical. However, the semi-empirical methods remain important for studying large molecules or large numbers of moleculesOne basic difference is that the ab initio results require both quantum and thermal corrections whereas, the semi-empirical calculations produce enthalpies of formation directly.

## 7.2.5. Ab Initio MO Theory

The ab initio methods have limited use due to the high cost involved in the method. In the simplest theory, which is termed as Hartree-Fock (HF) or self-consistent-field (SCF), one electron generates an average electric field in which the electrons move. Mathematically, the shape of the MOs is the linear combination of simpler function. This resembles the atomic orbitals and constitute the basis set.

If the size of the basis set is increased by adding more basis functions, it gives the better description of the detailed shape of the MOs. More reliable results are obtained from the bigger basis sets. However, the cost is high. The computational expenses of an H F calculation typically increase as N2.7. N is the number of basic functions (which is approximately proportional to the number of atoms).

The active area of research is the reducing dependence. Due to the instantaneous repulsion between electrons, termed electron correlation energy or simply correlation energy, there was an attempt to include the energy. However, more sophisticated theories begin with the HF calculations.

The perturbation theory (e.g., MP2, MP4), which is essentially a theory of extrapolation, by configuration interaction (CI) theory (e.g., CISD), or by coupled cluster theory (e.g., QCISD, CCSD (coupled-cluster, singles, and doubles)), helped in the inclusion of the correlation energy.

Like HF theory uses a basis set of atomic orbitals to describe each two-electron or one-electron, MO, correlated theories use a basis set of electron configurations (i.e., individual choices of which MO are occupied, and which are empty) to describe the many-electron wave function. Correlated calculations are much more accurate than uncorrelated HF calculations. There is one negative aspect related to this and it is the high cost associated with it.

Moreover, the correlated calculation requires a larger basis sets that the HF calculations.

Electrons correlation is essential in quantitative calculations of any process because most correlation energy is associated with electron pairs. The process should be such that it makes or breaks electrons pairs.

The electron correlation is important in the theoretical description of most chemical reactions, since most of the chemical bonds involve an electron pair.

Correlation energy may be included explicitly or implicitly within the theory either by means of empirical corrections or chemical strategies or as discussed before. For obtaining accurate results, both large basis sets and highly correlated theories are needed.

However, when the computational chemists are discussing highs and lows of ab initio theories, they mean to the sophistication of the treatment of correlation energy and then to the quality of the basis set.

The example of the above is, HF theory is the lowest level of theory and hence is uncorrelated. The second-order perturbation theory, i.e., MP2 theory is the lowest correlated level. The fourth-order theory, i.e., MP4 is the higher-level of theory than MP2.

However, the coupled-cluster theory (CCD (coupled-cluster, doubles) theory) that includes only the double-excitation operator is also higher than MP2. CCSD(T) (coupled-cluster, singles, and doubles and triples) is the theory that includes coupled-cluster theory plus the single-excitation operator and an approximate treatment of the triple-excitation operator. Since 1997, this theory is the highest level of theory in general use.

There are basis sets that carry similar names, in which bigger basis sets have longer designation or they include larger numbers. For example, 6–3 1 1 ++G(2df,2pd) is the bigger basis set whereas, 6–3 1 G(d) is the small basis set.

The ab initio calculations produce total molecular energies expressed in the atomic unit of energy and they do not produce enthalpies of formation or any other thermochemical quantities directly. The calculation of ab initio is expressed in atomic unit of energy, for example, the hartree (1 hartree /molecule = 627.51 kcal/mol = 2625.5 kJ/mol). The negative values of ab initio calculations represent the energy change upon assembling the molecule from its component nuclei and electrons.

The textbook methods of statistical mechanics are used in finding out the entropies, heat capacities, and other quantities derived from the computed molecular partition function. However, the only method for finding out the enthalpies of formation is computing the energy change for some chemical reaction. However, total atomization is a common choice of reaction for this purpose, rather it is optional.

## 7.2.6. Multireference MO Theories

The molecules which using single electron configuration are based on Hartree-Fock theory in the ab-initio method. There are some instances where there is more than one electron configuration, that is needed for correct qualitative description of the bonding. Such molecules are described by using multireference configuration SCF, MCSCF calculation, possibly followed by perturbation theory (e.g., CASPT2) or CI (e.g., MR-SDCI) to include additional electron correlation.

However, these methods can only be used by the specialists. The specialist has to take many decisions before applying these techniques and so far, there is no specific algorithm for making such decisions. In such cases, expert judgment is useful but sometimes specialists to make bad choices and hence poor results are obtained.

## 7.2.7. Density Functional Theory (DFT)

According to the fundamental theorem of DFT, there exists a functional, i.e., a function of a function of the electron density that yields the exact electronic energy. However, the theorem does not provide the correct functional. Also, it does not give any instructions for obtaining correct functions. The development of better functional is dependent on the significant progress in the DFT thermochemistry.

The theory does not have any higher order as in the case of MO theory. The chemists like Curtiss and Raghavachari, Politzer, and Seminario, Blomberg, and Siegbahn, and Ziegler has penned down the usefulness of the DFT in thermochemistry. DFT has its own advantage over MO theory. The highly correlated theories such as QCISD are comparable to the correlation energy. This is for a cost between those for HF and MP2 calculations.

By using smaller basis sets, the results obtained are usually converged than are needed for correlated MO methods. DFT is mostly useful for molecules of multiconfiguration character, free radicals with spin contamination (i.e., undesired mixing of high-spin excited states with the ground state), or molecules containing transition metal atoms. Therefore, it can be said that DFT methods are a robust system for difficult systems.

Disadvantages of DFT

- The main disadvantage of DFT for thermochemistry is the lack of extensibility. Extensibility here means that the theory cannot support the series of calculations to demonstrate convergence.

- There is no high level DFT calculation, however, non-local functional are fancy than the local functionals that are more reliable.

- One other disadvantage of the DFT is its general inability to cope with electronically excited states of molecules.

There exists a practical difference between DFT and MO theory. Computationally, the DFT is less expensive and provides thermochemistry of greater precision (fewer outliers) with lower accuracy in which larger typical errors are there.

## 7.2.8. Quantum Monte Carlo (QMC)

The QMC method is basically used to solve the Schrodinger equation. The ab initio calculations are done by few specialists in QMC methods. The calculations by this technique helps in achieving precisions proportional to the square root of the computational expense.

The QMC method has a unique feature that the sampling statistics defines most of the uncertainty of the result. For example, a calculation on hydrogen fluoride yielded a (non-relativistic) bond strength De = 592.2 ± 1.7 kJ/mol, only 0.12 kJ/mol greater than the corresponding experimental value.

One big disadvantage of the QMC Method is that it is very expensive that even at 10 electrons, the hydrogen fluoride is the biggest molecule that is feasible at that time.

However, approximations can be used to extend the range of QMC calculations. This adds to the uncertainty of the results and compromises the advantage of the technique.

Thus, QMC Methods are not yet practical for typical thermochemical applications.

Nearly ab initio Methods: Empirical Corrections

There is a probability that people will mostly rely on the calculation done by ab initio methods. The users of thermochemical data have been seen relying on the ab initio method used for predictions. However, there has been a growing consensus among people that the ab initio method alone cannot satisfy the demands for thermochemical data.

In order to resolve this mismatch, empirical corrections must be applied for the needed accuracy. It should be noted that, empirical corrections are feasible only to the extent that

(1)    The errors in the theory are systematic and predictable; and

(2)    In order to implement the corrections, there are enough experimental data.

## 7.2.9. Parameterized Methods

There are many methods existing that are parameterized and there is a continuous effort in developing the empirically parameterized solutions. To obtain an enthalpy of formation, there are some of the oldest method that involve atom equivalent that are the surrogates for atomic energies that are combined with the total molecular energy. These methods are not costly, easy to use and accurate.

Presently, the utility of this method is limited because this method can be best applied to common molecules for which there is already a method called, empirical group method.

Going forward, the methods and theories are expected to become more useful and advanced. At the other end, there are expensive end of the spectrum, Gaussian-2 (or G2) method and its variants which are popular among physical chemist.

Gaussian-2 (or G2) methods are the composite methods in which the results of several calculations are combined for estimating the results that would have been obtained with a much more costly calculation.

There are few empirical parameters that do not depend upon the chemical composition of the molecule under study, which results in a general method. These techniques again are easy to use and usually accurate (typically to 10 kJ/mol or so). The only negative aspect attached with this method is its high cost, which limits its use to smaller molecules.Furthermore, the composite methods occasionally give poor results. Also, semiempirical MO theories include empirical parameters at the quantum mechanics level, not as simple arithmetic corrections. There are some nominal ab initio approaches that do the same, although with fewer adjustable parameters.

The most popular method is B3LYP method. B3LYP (Becke, 3-Parameter, Lee–Yang–Parr) method is the hybrid, parametrized mix of local DFT, nonlocal DFT, and HF theories. This method is expensive than the HF theory and is easy to use and is more reliable. Since it includes correlation energy self-consistently.

The BAC-MP 4 method is the one that combines moderately high-level calculations. The combination is done with several empirical parameters based primarily upon the types and lengths of chemical bonds.

The BAC-MP 4 is one of the least expensive of all the established methods. This method is unique in itself as it provides an estimate of the uncertainty of each prediction. However, since the parameter development requires experimental benchmark data for each bond type, predictions for unusual types of molecule are unreliable. This method would have gained much more importance if the software were readily available to automate the tedious arithmetic involved. PCI-X (parametrized correlation method, containing a single adjustable parameter X) is such a method.

# 7.3. CHEMIST'S STRATEGIES: REACTION SCHEMES

The procedure is not uniquely determined by the molecule under study, where the most popular empirical corrections are unsystematic. It is based on the notion that the chemically similar molecule will have similar semantic errors having theoretical calculation. Then efforts are made to find out that which molecule is similar to the one of interest and for which the thermochemistry theory is readily available.

With much smaller basis sets, the result of the experiment using a reaction would be excellent. There would be need of a theory that is much lower level. One of the tests is for comparing its energy change calculated at the uncorrelated HF level with that calculated at a correlated level such as MP2.

When there is no effect of the explicitly including correlation energy to the reaction energy, then, there is well-balanced correlation effects in the reaction.

As earlier, it was made clear that the electron correlation effects in the reaction are very important in the energetics of reactions that involves a change in the number of electron pairs, or equally in the spin multiplicity.

The ab initio prediction has one disadvantage that it suffers from systematic errors for such processes.

However, it is also said that there is no practical calculation that can recover all the correlation energy. There is a process which is called Bond-breaking process.

For example, all the electrons in chlorobenzene are paired but one pair break to form C6H5 + Cl.

Hence, by simply calculating the energy for the below mentioned reaction would not result in reliable energy bond.

Simply,

$$C6H5Cl \rightarrow C6H5 + Cl \tag{1}$$

By using the popular unrestricted Hartree-Fock (HF) calculations, small 6–31G* basis set predict an enthalpy change.

$\Delta rH°298 = 230$ kJ/mol

The HF result is compared miserably with the experimental value of $396 \pm 8$ kJ/mol; the error is $-166 \pm 8$ kJ/mol. (Energies calculated using the hybrid B3LY P density functional method yield $\Delta rH°298 = -379$ kJ/mol, for a much smaller error of $-17 \pm 8$ kJ/mol.)

## 7.3.1. Isogyric Reactions

A reaction where the pair of electrons does not change is an isogyric reaction. In other words, an isogyric reaction is one in which the spin multiplicity does not change. This reaction is constructed by adding hydrogen atoms molecules to balance the spin multiplicity.

$$C6H5Cl + H + H \rightarrow C6H5 + Cl + H2. \tag{2}$$

The HF prediction for the isogyric reaction is

$$\Delta rH°298 = -98 \text{ kJ/mol}$$

As compared with the experimental value of $-40 \pm 8$ kJ/mol. Now the error is reduced to $-58 \pm 8$ kJ/mol. (B3LYP energies yield $\Delta rH°298 = -67$ kJ/mol).

Of course, one may choose any species to balance the spin. The reaction given below, i.e., reaction 3 is also isogyric and it is more concise. For reaction 3, HF was reacted with C6H5Cl:

$$C6H5Cl + HF \rightarrow C6H5F + HC1 \tag{3}$$

Prediction of $\Delta rH°298 = -63$ kJ/mol is compared reasonably well with the experimental value of -36 $\pm$ 8 kJ/mol; the error is $-27 \pm 8$ kJ/mol. (B3LY P energies yield $\Delta rH°298 - 32$ kJ/mol).

Finally, reaction 4 is isogyric and also avoids explicit calculations on free

$$C6H5Cl + H2 \rightarrow C6H6 + HC1 \tag{4}$$

Radicals, which are more often problematic than the calculations on closed-shell molecules are. For this reaction, the HF prediction of $\Delta rH°298 = -74$ kJ/mol agrees well with the experimental value of $-64 \pm 1$ kJ/mol; the error is only $-10 \pm 1$ kJ/mol. (B3LY P energies yield $\Delta rH°298 = -50$ kJ/mol).

Such that in reaction 2, the important "higher-level correction" in G2-type theories originated from considering processes. The calculations on H and H2 explicitly are unnecessary; the (constant) energy for the hypothetical reaction H $\rightarrow$ ½ H2 is applied once for each unpaired electron in the molecule.

## 7.3.2. Isodesmic Reactions

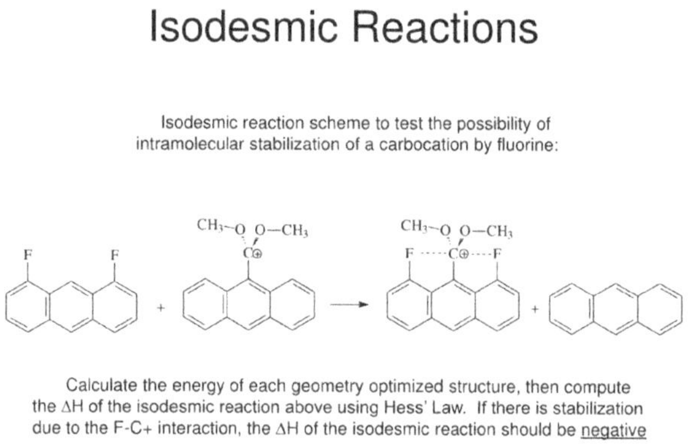

Isodesmic Reactions

Isodesmic reaction scheme to test the possibility of
intramolecular stabilization of a carbocation by fluorine:

Calculate the energy of each geometry optimized structure, then compute
the ΔH of the isodesmic reaction above using Hess' Law. If there is stabilization
due to the F-C+ interaction, the ΔH of the isodesmic reaction should be <u>negative</u>

Figure 7.2. Isodesmic reaction.

*Source: Image by slide serve.*

The reaction in which the number of chemical bonds of each formal type do not change. In other words, an isodesmic reaction is one in which the numbers of chemical bonds of each formal type, for, e.g., C-C, C=0, C-H) do not change.

For example, there are six aromatic C-C bonds, nine C- H bonds, and one C-Cl bond on each side of reaction 5. The inexpensive H F calculations predict $\Delta rH°298 = 10$ kJ/mol for reaction

$$C6H5Cl + CH4 \rightarrow C6H6 + CH3Cl \tag{5}$$

The experimental value is $21 \pm 1$ kJ/mol, for a small error of $-11 \pm 1$ kJ/mol. (B3LY P energies yield $\Delta rH°298 = 21$ kJ/mol). One may balance the chemistry at an even finer level of detail. Reaction 6 is isodesmic even if one distinguishes between sp3- and sp2-

$$C6H5Cl + C2H4 \rightarrow C6H6 + C2H3Cl$$

hybridized carbon atoms. In this case the calculated enthalpy changes of

$-3$ kJ/mol is in perfect agreement with the experimental $\Delta rH°298 = -3 \pm 2$ kJ/mol. (B3LY P energies also yield $\Delta rH°298 = -3$ kJ/mol)

All the ab initio calculations were performed on the personal computer for reactions 1–6 (geometry optimizations, vibrational frequency calculations, and energy evaluations). It has been said earlier also that, chemically balanced reaction schemes are useful not only for converting theoretical energies into enthalpies of formations, but it also helps in making purely empirical estimates.

For example, if there is one simply assumed reaction 6 to be thermoneutral, the corresponding prediction for $\Delta rH°298$ (C6H5C1) would be in error by only $-3 \pm 2$ kJ/mol. Therefore, in order to construct well-balanced reaction schemes, sometimes the experimental data are not available. Then, one must then rely on chemically similar functional groups, such as in isovalent homologous or congeneric series.

# 7.4. CURRENT CHALLENGES AND OUTLOOK

Thermochemistry involves many challenges in improving the state of computational thermochemistry. While, empirical estimation methods are still evolving and growing slowly. These methods are continued to be constrained by the experimental databases.

There are many potential users who finds it too expensive, too intimidating, too difficult, or too approximate to use the quantum chemistry, however, the quantum chemistry is found to be promising. The academic researchers have accepted these challenges primarily. Technological issues such as improving the quality of experimental databases and computer software are being addressed in the industrial and government sectors.

Although the methods are good to use, there exists the cultural problem of acceptance and using the newer methods. The reliability of the new methods is questioned.

## 7.4.1. Computational Technology

The speed of the computer is increasing rapidly. Raw computing power of the computer can be used to a greater extent, so the algorithms are constantly being improved. Resulting to which, the cost of a given computational prediction is declining very quickly.

In contrast to this, the experimental thermochemistry is getting costly and the number of well-equipped laboratories is getting stagnant and declining. Therefore, it can be concluded that in thermochemistry, computational approaches will become increasingly important.

Also, the computer will constantly begin to improve and have a robust system and also it will be easier to use. This is a fact that is driven by the emergence of the commercial vendors of software for quantum chemistry. Since there is limited time for training and development, customers expect computers to have such software that can be accessible by any non-specialist also.

The demand for the computer programs is increasing due to the demand for user-friendly software throughout the software industry. The convenient graphical interface is what the customers expect from the dealers.

## 7.4.2. Scientific Developments

For the prediction in thermochemistry, there are research in the methods of MO theory and the DFT theory in many directions. The development of improved algorithms is being there for efficient implementation of the existing theories. This will allow the study of the larger molecule. The new basis sets extend current methods to additional chemical elements.

The improvement in the functional increase the reliability of DFT predictions. The reliability of the MO theory is improved by the sophisticated theories of electron correlation. The quantitative information about the reliability and the range of applicability of new theories and methods can be found by the widespread testing.

There is a need to have better theories of molecular solvation If the quantum chemistry is to be trusted for predicting thermochemistry in solution and not just in the gas phase. The advances that has been there in the ab initio method are also applicable to semiempirical method.

There has been development in the underlying semiempirical theory with the development of rigorous Hamiltonians that led to more accurate results. There are less-developed areas of theory that will affect computational thermochemistry as they grow.

In the high-level MO theory, it has some trouble with some difficult molecules. However, the multireference theory did well with these problems but it depends upon the user's judgment and hence an expert user is required. The efforts were made to search for the automatic ways to make necessary choices, but not enough to be used by non-specialists and others.

Another emerging area are the vibrational zero-point energy and also the molecular partition functions that involves efficient, sophisticated treatments of molecular vibration. The errors found in the ab initio thermochemistry are currently assumed to be in a molecule's electronic energy. The calculations of electronic energies are becoming very accurate. However, the accuracy of predictions, especially for floppy molecules is limited in the vibrational energy.

## 7.4.3. Symposium Panel Discussion

There was a panel discussion that was held as part of the symposium. The topic on which the discussion was done was the needs for and applications of accurate thermochemical data that emphasized mainly on prediction.

In the discussion, the panel members represented the users of thermochemical data in the areas of microelectronics, chemical processing, academia, and the military. Common problems and goals were discussed by the users and the developers in an open forum. The discussion has reached four major conclusions, which are as follow:

*Benchmark Data*

There is a need for experimental database that is critically evaluated in order to provide a standard metric for evaluating, comparing, and parametrizing the performance of different predictive methods. This was the main topic over which the discussion was done, and the participants thought would help both the developers and users of new technique.

Today, mostly every developer of the technique has their own databases, which is used for data compilation. A new database was made for public use by Curtiss (Argonne NL) that was compiled for testing the G2 method and its variants.

Wider Range of Properties and Molecular Types

There must be accurate predictions for engineering properties and also for a wider variety of molecular types. Where the thermochemical properties are important in industries, there are many more properties like kinetics, vapor-liquid equilibria, critical properties which are necessary for making a process economically successful.

The discussion also pointed out that work is being done on other properties also, some papers on the topic were presented at the symposium. The application of predictive methods to organometallic compounds and to the properties of molecules on surfaces is also important.

*Acceptance among Engineers*

There are broader thermochemistry community that thinks to be unaware of the new predictive techniques that exist and are beneficial. Many companies like DuPont, Dow have strong molecular modelling department, but they don't know how beneficial the techniques are.

The continuing efforts and education are needed to achieve this symposium. The techniques must be easy to use in order to gain widespread acceptance and use.

### Hidden Experimental Data

The heavy parametrized methods are limited by the available experimental data. However, the experimental measurements are done in industrial labs. Nowadays, most of the corporations have clearance processes and they do not encourage the dissemination of private information. Although, keeping data private sometimes provides a competitive advantage, but this is not the case.

## 7.5. THERMODYNAMICS

The study of energy interactions between systems and the effect of these interactions on the system properties is known as Thermodynamics. There is a transfer of energy between systems that takes place in the form of heat or work. Thermodynamics is a subject that deals with system equilibrium.

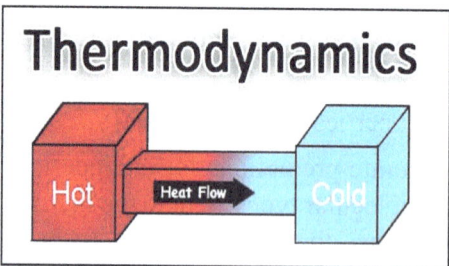

Figure 7.3. Thermodynamics is concerned with heat transfer.

*Source: Image by Jagran josh.*

A thermodynamic system is defined as the study of quantity of fixed mass and identity upon which attention is focused. Thermodynamic system is the everything this chapter needs to study.

In thermodynamics, a system could be as simple as gas in a cylinder and as complex as a nuclear power plant. Things that are external to the system is the surroundings. There is a boundary that separates system from its surroundings, the boundary is known as system boundaries.

The thermodynamic system is classified into three different systems, i.e., closed systems, open systems and isolated systems. There is a specified region in space where most of the attention is focused, is known as control volume. The control volume can be considered as an open system. The control surface separates the surroundings from control volume. This results in easy entry and exit of the mass and energy in the control room. There involves a wide variety of processes in refrigerating and air conditioning. The understanding of the nature of the process path is important since heat and work depends on the path. In this, the process can be said to have different states that occurs within the initial state to the final state. In thermodynamics, the process is said to have completed a cycle if the process starts from the initial state and finally arrives at the initial state.

According to the website of the Energy Education of the Texas Education Agency, Thermal energy is the energy a substance or system has due to its temperature. A professor of physics at Missouri Southern State University, David McKee has defined thermodynamics that involves measuring this energy, which can be "exceedingly complicated." Because there are large numbers of atoms and molecules that interact in a complicated way.

These atoms and molecules can be described with a very small number of measurements or numbers if equilibrium is achieved. This is known to be the mass of the system, the pressure of the system, and the volume of the system, or some other equivalent set of numbers.

## 7.6. RELATION

The branch of physical science that deals with the relationship between heat and other forms of energy such as mechanical, electrical, or chemical energy is known as thermodynamics. Whereas, Thermochemistry is a branch of thermodynamics. Thermochemistry comes under thermodynamics. Thermochemistry is also a branch of chemistry that describes the heat energy in relation to chemical reactions.

There is a key difference between computational thermochemistry and thermodynamics is that the computational thermochemistry is the quantitative study of the relation between heat and chemical reactions which is done on computers. Whereas, thermodynamics is the study of laws associated with the relation between heat and chemical reactions.

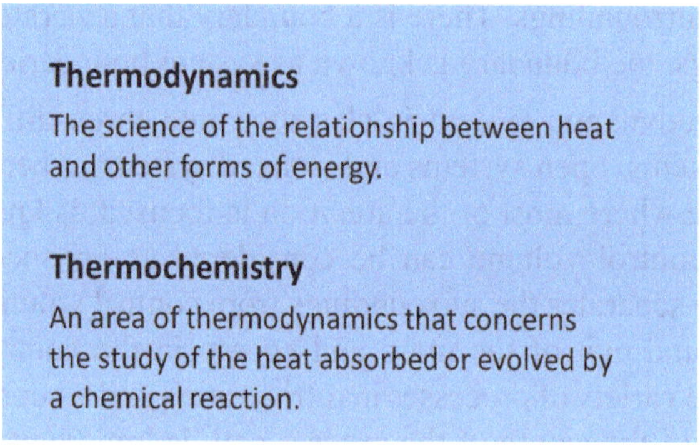

> **Thermodynamics**
> The science of the relationship between heat and other forms of energy.
>
> **Thermochemistry**
> An area of thermodynamics that concerns the study of the heat absorbed or evolved by a chemical reaction.

Figure 7.4. Thermodynamics vs thermochemistry.

*Source: Image by slide player.*

The calculation involved in the computational thermochemistry is performed on the computers, while thermodynamics deals with heat and energy.

In thermochemistry, there is a term that is often used known as "enthalpy." Enthalpy is the thermodynamic quantity equivalent to the total heat content of a system. In mathematical words, the enthalpy ($\Delta H$) is equal to the internal energy of the system plus the product of pressure (P) and volume (V).

The formula for enthalpy can be written mathematically as:

$\Delta H = U + PV$

By using the enthalpies of different chemical species, the determination of the heat of reaction and many other parameters is done. The heat of reaction is the change in enthalpy. It is given by the difference between the enthalpy of products and enthalpy of reactants. $\Delta H = \Delta H$ (products) – $\Delta H$(reactants).

Computational thermochemistry describes the relationship between heat energy and chemical reactions. Whereas, Thermodynamics describes the relationship between all energy forms with that of heat energy.

The word thermodynamics is derived from the Greek words that mean heat and power. Thermodynamics is studied and has applications in all the sciences. Whereas, thermochemistry is the part of thermodynamics that studies the relationship between heat and chemical reactions. Thermochemistry helps in determining if the particular reaction will occur and if it will release or absorb energy as it occurs. That is why it is considered as the very important field of study.

Thermodynamics also helps in calculating how much energy a reaction will release or absorb. This information can be used to determine if it is economically viable to use a particular chemical process. In thermochemistry, it does not predict how fast a reaction will occur.

Table 7.2. Comparison Between Computational Thermochemistry and Thermodynamics

| Computational Thermochemistry | Thermodynamics |
|---|---|
| Thermochemistry is also a branch of chemistry that describes the heat energy in relation to chemical reactions. | branch of physical science that deals with the relationship between heat and other forms of energy |
| Computational thermochemistry describes the relationship between heat energy and chemical reactions | Thermodynamics describes the relationship between all energy forms with that of heat energy |
| In thermochemistry, it does not predict how fast a reaction will occur | Thermodynamics also helps in calculating how much energy a reaction will release or absorb |

In the view of world, terminology of thermochemistry should be defined in terms of thermodynamics. The chemical reaction being studied is considered to be the "system." For example, if there is a testing being performed, an acid is being mixed with a base, the acid, the base, any H2O used to dissolve them and the beaker in which the reaction is performed are considered the system.

Everything that is not the part of the system is considered to be its "surroundings." Surrounding includes everything from the countertop on which the beaker is held to the planets in outer space. The system and surroundings together form the "universe."

From the above discussions, it is easy to understand the relation between computational thermochemistry and thermodynamics.

## 7.7. CONCLUSION

Thermochemistry is the study of the heat released or absorbed as a result of chemical reactions. It is characterized by systems (open, closed, and isolated) and processes (isothermal, adiabatic, isobaric, and isovolumetric). When the thermochemical predictions are made using computer systems, the thermochemistry is then called computational thermochemistry. The major importance of thermochemistry of molecules in the chemical sciences and is essential to many technologies. With the advancement in the system nowadays, it is possible to compute reliable thermochemical data for a fairly wide variety of molecules through theoretical methodology, computer algorithms, and computer power

Also, the computational thermochemistry describes the Gaussian-n series of methods that have been widely used for thermochemical predictions. There are challenges in using computational thermochemistry and the steps taken would be the extension of the methods to larger molecules, increased accuracy in predictions, and extension to heavier elements.

The different methods in computational thermochemistry has been discussed and in the end the relation between thermodynamics and the computational thermochemistry is explained.

# REFERENCES

1.   Chemistryexplained.com. (n.d.). Thermochemistry – Chemistry Encyclopedia – Reaction, Water, Gas, Symbol, Equation, Mass, Energy and Enthalpy. [online] Available at: http://www.chemistryexplained. com/Te-Va/Thermochemistry.html (accessed on 13 March 2020).

2.   Curtiss, L., & Pople, J., (n.d.). Recent Advances in Computational Thermochemistry and Challenges for the Future. [online] nap. Available at: https://www.nap.edu/read/9591/chapter/5#33 (accessed on 13 March 2020).

3.   Difference Between Thermochemistry and Thermodynamics, (n.d.). [ebook] Available at: https://www.differencebetween.com/wp-content/uploads/2018/02/Difference-Between-Thermochemistry-and-Thermodynamics.pdf (accessed on 13 March 2020).

4.   Partington, J., (1964). Thermochemistry and Thermodynamics. [online] springer. Available at: https://link.springer.com/chapter/10.1007/978–1-349–00554–3_19 (accessed on 13 March 2020).

5.   science. jrank, (n.d.). Thermochemistry. [online] Available at: https://science.jrank.org/pages/6793/Thermochemistry-Thermodynamics-thermochemistry.html (accessed on 13 March 2020).

6.   Science.jrank.org. (n.d.). Thermochemistry – Change. [online] Available at: https://science.jrank.org/pages/6794/Thermochemistry-Change.html (accessed on 13 March 2020).

7.   Warn, J., & Peters, A., (n.d.). Concise Chemical Thermodynamics. [ebook] Available at: https://www.mobt3ath.com/uplode/book/book-36809.pdf (accessed on 13 March 2020).

# Applications of Thermochemistry

## CONTENTS

Thermochemistry, which is concerned with the absorption and release of heat energy during chemical reactions and physical process has many useful applications. It helps determine if a reaction will release or absorb energy and how much energy will be absorbed or released.

This helps in developing many applications using chemical process and test the viability of these processes. This chapter discusses some important current applications of thermochemistry as well as some techniques which are still emerging and with further research can be used on a large scale.

This chapter first describes the applications of enthalpy and its uses in calculating energy content of fuels. A section as also attributed to explaining how understanding of thermochemistry is applied in the field of nutrition. The chapter also describes isothermal titration calorimetry (ITC), an important application of thermochemistry which measures the heat exchange that takes place when two molecules bind. It is used in characterizing binding affinity of ligands for proteins which is has its uses in drug discovery. There are a few emerging techniques designed based on the principles of thermochemistry which are also tackled and explained in this chapter. These include thermochemical conversion of lignocellulosic biomass for liquid fuels which is aimed to address the problem of non-renewable sources of energy. Thermochemical splitting of water is also being investigated for producing hydrogen while minimizing the release of greenhouse gases. Finally, thermochemical energy storage technology is discussed in detail.

## 8.1. APPLICATIONS OF ENTHALPY

Majority of the chemical reactions that take place, involves the absorption or release of heat. This has an impact on the internal energy of the system. Enthalpy and the First Law of Thermodynamics have several chemistry related applications. Its application is not limited to chemistry alone, but also in everyday life.

### 8.1.1. Enthalpy Diagrams and Their Uses

Enthalpy changes can be explained by graphical construction in which the relative enthalpies of various substances are represented by horizontal lines on a vertical energy scale. Energies are always arbitrary and hence the zero scale can be placed anywhere. Elements are generally located at zero energy, since it reflects the convention that their standard enthalpies of formation are zero.

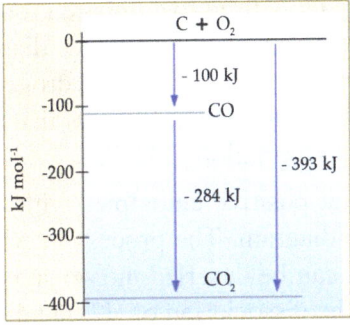

**Figure 8.1.** Enthalpy of CO

The system undergoes various reactions which causes changes in enthalpy. The above diagram shows these changes. The enthalpy changes for the combustion of carbon monoxide to carbon dioxide (CO2) can be derived by subtracting appropriate arrow lengths. The thermochemical equations need not be written in a formal way to calculate this.

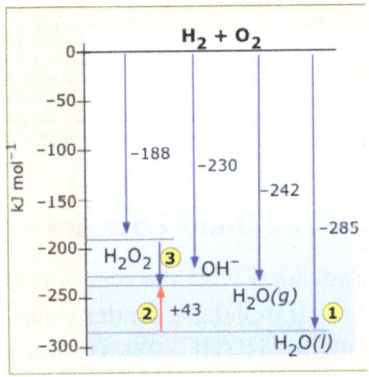

**Figure 8.2.** Enthalpy Diagram of H2O

The above diagram illustrates the known stable configurations of these two elements. When H2 and O2 react, it yields one mole of liquid water (H2O) and releases 285 kJ of heat. When the release happens in gaseous state, the energy released will be smaller.

The above diagram clearly shows the heat of vaporization of H2O. Spontaneous decomposition of Hydrogen peroxide forms H2O and O2 during which heat is released. Generally, these reactions are slow, so the heat is not noticed. However, when an appropriate catalyst is used, the reactions can become fast and can even be used to fuel a racing car.

Enthalpy diagrams are widely used to compare groups of substances which have some common features. The following diagram shows the molar enthalpies of species with respect to two hydrogen halides belonging to those elements. The following diagram shows that the formation of HF is more exothermic than the formation of HCl.

Gaseous atoms are at positive enthalpies with respect to the elements in the upper part of the diagram. The process in which the H2 and the di-halogen are dissociated can be descried in two steps. The enthalpy change that takes place during the dissociation of H2, the dissociation enthalpies of F2 and Cl2 can be calculated and shown in the diagram in Figure 8.3.

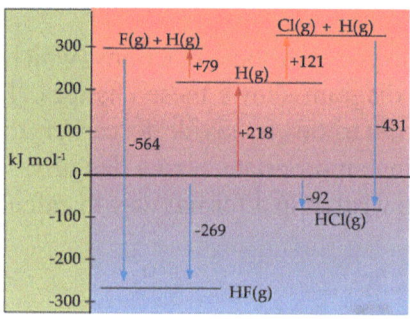

**Figure 8.3**. Enthalpy Diagram some halogens and their acids.

## 8.1.2. Bond Enthalpies vs. Bond Energies

The enthalpy change associated with the reaction $HI(g) \rightarrow H(g) + I(g)$ is the enthalpy of dissociation of HI molecule. Under standard conditions, it can be expressed as the bond enthalpy H°(HI, 298K) or internal energy U°(HI,298). The difference between the two quantities in this case is measured as $\Delta PV = RT$. This reaction cannot be studied directly, and hence appropriate standard enthalpies of formation is used.

| | |
|---|---|
| ½ H2(g)→ H(g) | +218 kJ |
| ½ I2(g)→ I(g) | +107 kJ |
| ½ H2(g) + ½I2(g)→ HI(g) | −36 kJ |
| HI(g) → H(g) + I(g) | +299 kJ |

Bond energies and enthalpies are important properties of chemical bonds, and it is very important to be able to estimate their values from other thermochemical data. The total bond enthalpy of a more complex molecule such as ethane can be found from the following combination of reactions:

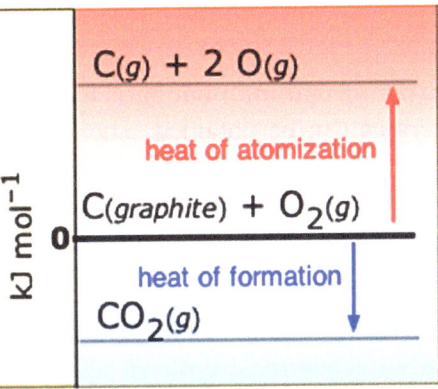

**Figure 8.3.** Comparing Standard Heats of Atomization (red) and Formation (blue) for carbon

Molecules can be broken into gaseous atoms from their ordinary state. This process is called atomization. In atomization process, heat is always absorbed. The heats of atomization are helpful in calculating bond energies. The measurement is done spectroscopically.

### 8.1.3. Pauling's Rule and Average Bond Energy

The Pauling's rule states that the total bond energy of a molecule can be thought of as the sum of the energies of the individual bonds.

This rule is only an approximation. This is because the energy of any given bond is influenced by the particular chemical environment of the two atoms and is not constant. Hence, the average energy of a particular kind of bond can only be discussed. For example, in case of C–O, the average is taken over a representative sample of compounds like CO, CO2, COCl2, etc. which contains this type of bond.

The bond energies do not have strict additivity. Pauling's Rule is still useful because it is useful in estimating heats of formation of compounds which have not be studied or prepared. In the example given below, if the enthalpies of the C–C and C–H bonds from other data can be known, the total bond enthalpy of ethane can be estimated. This can be back calculated to derive some other quantity of interest such as ethane's enthalpy of formation.

When a large amount of experimental information of similar nature is assembled, a consistent set of average bond energies can be derived. When a large amount of experimental information is assembled, a consistent set of average bond energies can be obtained. The energies of double bonds are higher than single bonds. The energies of triple bonds are greater than double bond.

## 8.1.4. Energy Content of Fuels

A fuel is a product which provides useful amounts of energy, through a controlled process at an economical cost. Generally, combustion in air is used as O2 is freely available in the atmosphere. The enthalpy of combustion helps determine a substance's suitability as a fuel.

There are other criteria as well, such as ignition property of the substance. Fuel, which is used for self-powered vehicles, should have a large energy density in terms of both mass and volume. Therefore, substances such as methane and propane must be stored as pressurized liquids while transporting it and other portable applications.

**Table 8.1.** Energy Densities of Some Common Fuels

| Fuel | MJ kg-1 |
|---|---|
| Wood (dry) | 15 |
| Coal (poor) | 15 |
| Coal (premium) | 27 |
| Ethanol | 30 |
| Petroleum-derived products | 45 |
| Methane, liquified natural gas | 54 |
| Hydrogen | 140 |

*Source: Table by Chem.libretexts.org.*

The U.S. agricultural industry is promoting Ethanol as a motor fuel. Some researches reveal that 46 MJ of energy is needed to produce 1 kg of ethanol from corn. However, there are other studies which consider optimal farming techniques and by-product uses while calculating this figure and hence the figure varies.

Hydrogen has an extraordinarily high density due to its molar mass and high heat combustion. It would be an ideal fuel if its critical temperature was higher. Since there are many advantages of using hydrogen as a fuel, many studies have been conducted to get a large amount of H2 into a small volume of space.

Compressing the gas to a very high pressure is not a viable option, since the weight of the heavy-walled steel vessel required to bear the pressure would cause an unacceptable increase of the effective weight of the fuel. One technique holds some potential which exploits the ability of H2 to "dissolve" in certain transition metals. Heating helps in recovering hydrogen from the solid solution.

## 8.2. APPLICATION OF THERMOCHEMISTRY IN NUTRITION

Thermochemistry is also linked to biochemical processes in living beings. It is also used in the field of nutrition to calculate the calorific value of food. Thermochemistry explains how food supplies the energy which keeps the body cells functioning.

Body releases 80% of this energy to maintain the body temperature at a level, because of which a living being continues to survive.

Thermochemistry helps determine the calorific content of food. Enthalpy of combustion ($\Delta H comb$) per gram determines this value. The general reaction is:

food + excess $O_2(g) \rightarrow CO_2(g) + H_2O(l) + N_2(g)$

Different foods have different chemical composition because its caloric content varies. For example, sugar such as glucose produces 15.6 kJ/g during combustion, while a fatty acid like palmitic acid produces about 39 kJ/g.

Nutritionists study these fatty acids and sugars which are the basis of fats and carbohydrates. They fulfil a major part of the energy requirement of a living being. The average values assigned to these to building blocks are 38 kJ/g (about 9 Cal/g) and 17 kJ/g (about 4 Cal/g) respectively. However, differences in composition of these foods cause these values to vary.

Proteins, another major source of calories also vary as well. Proteins are primarily composed of amino acids with the following structure:

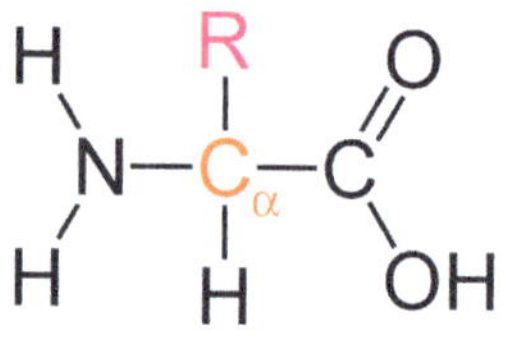

**Figure 8.4.** General Structure of amino acids.

*Source: Image by Wikimedia Commons.*

An amino acid is mainly composed of an amine group ($-NH2$) and a carboxylic acid group ($-CO2H$) besides other functional groups. The human body requires 10 amino acids from the food, since the body cannot synthesize them from other compounds. R can be any of the different groups of compound and hence each amino acid has a different value of $\Delta$Hcomb. The average value of $\Delta$Hcomb in proteins is 17 kJ/g (about 4 Cal/g).

The caloric content of food that are reported excludes $\Delta$Hcomb for those components that are not digested, such as fibre. Food substances such as meat and fruits contain 50% – 70% $H2O$ content. These cannot be oxidized by the body to release energy. Hence, the calorie value of $H2O$ is zero.

Some foods like broccoli and peas have a large amount of fibre which is mainly glucose. Though fibre can be burned in a calorimeter, humans cannot break it down to smaller molecules for oxidation. Hence, fibre is not taken into account while calculating the calorific content.Caloric content is calculated in two ways. The first method is to dry a carefully precise weight by sample and carry out combustion reaction in a bomb calorimeter. The other approach, which is commonly used is to analyse the composition of food and calculate the caloric content using average values for each of them. This is the approach commonly used by nutritionists while preparing a diet.

Calorie is a significant portion of energy. It is estimated that a person weighing 160 lb. needs 67 Cal/h in order to provide energy for the basic biochemical processes that is integral for their sustenance.

This energy maintains body temperature, heartbeat rate, facilitate chemical reactions in the body, fuel the muscles for breathing, sends nerve impulses which regulate the body functions. Physical activity expends energy and hence, people desirous of burning calories are advised increased physical activity.

An average individual who is moderately active require 2500–3000 Cal/day. Sportspersons and athletes can burn up to 4000 Cal/day. The excess calories which is not burned is converted into fat and used by the body for future use.

These stored fuels are mobilized and oxidized when the energy requirement exceed the energy supplied by diet. The supply of stored carbohydrates is tapped by the body first, before the fats. Hence, low-carbohydrate diets have become popular.

Different physical activities entail various levels of energy consumption. For example, a person needs more amount of energy to climb a hill than to walk. Also, the rate at which a person is doing various activities is a relevant factor in calculating the change in potential energy.

**Figure 8.5.** The calorie burnt in various physical activities are different.

*Source: Image by 59th Medical wing.*

These complex calculations like the net calories burned can be calculated by calculating the calories in a diet based on the food composition and the energy expended in these activities. Principles of thermochemistry are useful in these calculations.

# 8.3. ISOTHERMAL TITRATION CALORIMETRY

ITC has emerged as a biophysical method to characterize the interaction between label-free biomolecules in solution. Macromolecular interactions are very important cellular events. They play an important role in transduction pathways.

Therefore, extensive research has been carried out on macromolecular interactions. This has facilitated a deeper understanding of molecular mechanism which is the basis for physiological and pathophysiological processes. This has helped in designing useful drugs which can modulate macromolecular binding events which can lead to diseases.

ITC is a significant tool which characterizes molecular interactions. ITC measures heat exchanges that occur when two molecules bind. When molecular interactions take place, heat may be absorbed or released. The former is called endothermic reaction and latter is called exothermic reaction.

ITC determines the differential power which the heaters of the instrument to both the reference and sample cells provide. This helps in monitoring heat changes. This is important for counteracting any temperature difference between the two cells during the binding reaction in order to ensure that no temperature difference arises between the reference and the sample cells.

## 8.3.1. Measurement Principle

ITC has a sensitive calorimeter which measures the heat absorbed or released when titration of the ligand into the sample cell which contains biomolecule of interest, takes place gradually into the sample cell.

## 8.3.2. How It Works?

The microcalorimeter contains two cells, one is the reference cell containing H2O and the other one has the sample.

The reference cell contains Milli-Q H2O, a sample cell which contains a biomolecule and an automated injection syringe. The automated syringe contains the other binding molecule (ligand) which helps in titration of ligand into the sample cell.

An adiabatic jacket surrounds the sample and reference cell. The system can detect the temperature difference the reference and the sample cells. It has two heaters which supplies power to both reference and sample cells to maintain an absence of temperature difference between them, i.e., ($\Delta T = 0$).

The function of the microcalorimeter is to maintain these two cells at the same temperature. The calorimeter has a heat sensing device which detects the temperature difference and gives the feedback to its heaters which then balances this difference.

## 8.3.3. Making a Measurement

Both the reference and sample cell are set to the desired temperature. A syringe is used to load the ligand and inject into the sample device which has the protein which has to be tested. The protein solution is injected with a series of small aliquots of ligand. The calorimeter is extremely sensitive and can even detect and measure heat changes of a few millionths of a degree Celsius.

The ITC starts measuring all heat released as soon as the first injection is made, until the binding reaction reaches equilibrium. The amount of binding determines the quantity of heat measured.

The output of the instrument is measured in terms of $\mu$cal/sec and is the power required to maintain $\Delta T = 0$ between the reference and sample cells. The ligand-biomolecule binding induces temperature difference between the reference and sample cell. This is converted into the power that is required to bring both of them back to the same temperature during the binding reaction.

Further into the titration process, the ligand saturates the biomolecule in the sample cell. This leads to reduced interaction and therefore reduced heat exchange occurs till the time the biomolecule is fully saturated. At this point the instrument detects only that heat exchange which occurs due to dilution of ligand.

### 8.3.4. Results and Data Analysis

If the reaction is exothermic, the temperature of the sample cells becomes higher than the reference cell. This causes a downward peak in the signal. The signal assumes its initial position when the temperature of the two cells become equal.

Then the second small of the ligand is inserted into the sample cell. The ITC again detects this small heat change and compensates for it. A series of ligand injections are used to increase the molar ratio between the ligand and protein. As the protein starts to get saturated due to these injections, less binding of the ligand takes place. The heat change starts to decrease, and a point comes where the sample cells contains more ligand compared to protein. As a result, the reaction is brought to saturation.

It is important to measure heat transfer which occurs during bindings. This helps to determine the binding constants (KD), reaction stoichiometry (n), enthalpy ($\Delta H$) and entropy ($\Delta S$ accurately. Therefore, a complete thermodynamic profile of molecular interaction can be obtained. ITC is not just concerned affinities but can also explain the mechanisms which form the basis for molecular interaction. The knowledge gained about structure-function relationships helps in taking correct decisions in lead optimization and hit selection. In the graph below, an example of 1:1 binding reaction is shown. The area of each peak is integrated and plotted against the molar ratio of ligand to protein. The outcome is an isotherm, which can be fitted to a binding model and the affinity (KD) is derived from it. The stoichiometry is derived from the molar ratio which is at the centre of the binding.

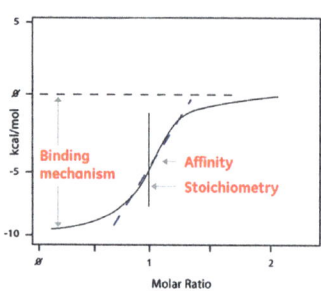

**Figure 8.6.** An example of a binding reaction diagram.

The enthalpy ($\Delta H$) can also be derived from isotherm. It refers to the amount of heat exchange per Mole of ligand bond. Even one ITC experiment can provide a wide range of knowledge about binding reactions. This helps in understanding how interaction takes place and the thermodynamic drivers behind it.

ITC has helps in drug discovery and development of many aspects some of which are outlined below:

- Selection and optimization of candidates;
- Measuring thermodynamics and active concentration;
- Quantifying binding affinity;
- Confirming intended binding targets in small molecule drug discovery;
- Validating IC50 and EC50 values during hit-to-lead;
- Measuring of enzyme kinetics;
- Characterizing the mechanism of action;
- Determining the binding specificity and stoichiometry; and
- The advantage of ITC is that it can simultaneously determine all binding parameters in a single experiment and is the is the only technique to do so. It does not require any modification of binding partners which may be required in other experiments

such as fluorescent tags or immobilization. The affinity of the binding partners is their native states can be measured with ITC.

# 8.4. DIRECT THERMOCHEMICAL CONVERSION OF LIGNOCELLULOSIC BIOMASS FOR LIQUID FUELS

There is an increased demand for liquid transportation fuels, while on the other hand, depletion of petroleum resources is occurring. These are raising environmental concern. The concerns can be addressed by the development of efficient conversion technologies which can enable the production of second-generation bio-fuels and non-food sources.

Thermochemical approaches have great potential for converting lignocellulosic biomass into liquid fuels. Thermochemical processes can enable this conversion in one step by making use of heat and other catalysts. These processes are more advantageous compared to indirect and biological processes. Some of them are integrated conversion of the entire biomass, greater flexibility of feedstock and lower costs of operation.

Thermochemistry plays and important role in the production of second-generation liquid biofuels from non-food biomass. Many direct thermochemical processes are used to produce liquid biofuels some of which are fast pyrolysis/liquefaction of lignocellulosic biomass for bio-oil, including upgrading methods, such as catalytic cracking and hydrogenation.

At the same time, biomass feedback involves reactions with catalyst/ chemical reactants which convert basic molecular structures into structures which are comparable with usable fuels.

The molecular structure change that happens during the thermochemical process is an important factor which indicates the quality of biofuels. The products that are derived from the biomass are compared against fossil fuels and a benchmark is established.

## 8.4.1. Thermochemical Processing of Lignocellulosic Biomass

This process generally involves the use an upgrading process of pyrolytic oil, the reason being the fuel properties of these oils are low when compared to petroleum fuels.

The upgrading process can be facilitated by thermochemical reactions. An example is the production of biochar from the process using dehydration reaction and condensation of liquid products. Majorly, two thermochemical processes are used for converting lignocellulosic biomass into liquid fuels namely pyrolysis and liquefaction. They are discussed below, but before that it is important to understand the processing properties of lignocellulosic biomass.

Sufficient lignocellulosic materials are derived from lumber production since 0% of timber is derived from solid waste. These wastes can be easily used for energy generation. Lignocellulosic biomass possesses different properties in comparison to petroleum-based fuels. It generally comprises of cellulose, hemicellulose, and lignin, in the proportions 35%–50%, 15%–25% and 15%–30%, respectively.

The oxygen present in lignocellulosic biomass varies between 40 to 50%. The source of this oxygen is bonded oxygen in cellulose and lignin. Lignin is composed of phenolic units, while cellulose is composed of linear chains of D-glucose units which are linked by β-(1–4)-glycosidic bonds.

**Figure 8.7.** The demand for biomass renewable energy is increasing.

*Source: Image by Pixabay.*

Therefore, a major challenge in attaining an elemental composition which is close to fossil fuels requires the removal of oxygen atoms. Thermochemical conversion facilitates the deoxygenation reactions of lignocellulosic biomass.

The fossil fuels are derived from the biomass feedstock which were formed centuries ago. Inside the earth, these biomass materials go through deoxygenation reactions, at a high temperature and pressure. The conversion of molecular structures of biomass change to petroleum crude oil is a slow process.

Thermochemistry is concerned with the relationship between the heat and pressure and its impact on the molecular structures in an attempt to mimic the reactions take place and produce petroleum oil.

## 8.4.2. Pyrolysis of Lignocellulosic Biomass for Liquid Fuels

1.    Fast Pyrolysis

Fast pyrolysis is method that is used to convert solid lignocellulosic biomass to a liquid product which is called pyrolytic oil or bio-oil. Fast pyrolysis conditions are used to generate bio-oil from whole biomass. A temperature of around 500°C, vapor phase temperature of 400–450°C is maintained. Short vapor residence is typically less than 2. A yield of around 75% of weight on a dry feed basis can be derived.

Generally, fast pyrolysis can generate biofuel from all forms of biomass. Researches have been conducted on several types of biomass from agricultural and forestry wastes such as sawdust and rice husk.

The bio-oil, which is derived through this process, contains:

- A mixture of multiple oxygenated components derived from depolymerization and fragmentation of cellulose.
- Hemicellulose and lignin, mainly composed of aldehydes, phenols, guaiacols, carboxylic acids, alcohols, esters, ketones, sugars, syringols, furans, lignin derived phenols.
- Extractible organics with multi-functional groups.

2.    Hydrogenation Upgrading

Thermodynamic equilibrium is not maintained for producing bio-oil. It involves a short reactor time which takes place at a high pyrolysis temperature. After this, it is rapidly cooled or quenched. This has a condensation effect, which is also not at thermodynamic equilibrium at storage temperature.

However, the chemical composition of the bio-oil tends to change towards thermodynamic equilibrium. Hence, it becomes important to upgrade processes for bio-oil production so the stability and fuel properties can be improved.

An effective technique for upgrading is the process of hydrodeoxygenation (HDO). This process is generally done at a moderate temperature of 300 – 600°C at which the oxygen is removed from original bio-oil. Catalysts and hydrogens are present during the process. Oxygen combines with hydrogen to form $H_2O$ and is removed in this manner.

The HDO process involves saturation of unsaturated components from different components of biomass. The pyrolytic products generated from cellulose have a specific reaction pathway which helps produce high grade biofuels (mainly short chain alkanes and alkenes). However, the phenolics from lignin tend form cycloalkane-rich fuels.

3.    Catalytic Cracking

Catalytic cracking is helpful in upgrading bio-oils. It not only reduces the oxygen content but also improves its thermal stability. The main advantage of catalytic cracking is that H2 is not required.

The operating cost is also lowered when atmospheric pressure processing is used. The temperature in which the process takes place is similar to those which is used to produce bio-oil. Hence, it is much more economic than HDO process. The only drawback is that poor yields of hydrocarbons and high yields of coke may take place in case of biomass-derived feedstocks.

When appropriate catalyst is used and appropriate conditions are maintained, the results can be improved.

When biomass-derived molecules are processes using catalytic cracking, the products include aliphatic and aromatic hydrocarbons, H2O-soluble organics, oil-soluble organics, H2O, gases such as CO2,CO and light alkanes and coke.

Originally, bio-oil was added into the reactor, along with the carrier gas by metering a pump through a preheater.

In most cases, the carrier gas is N2. The bio-oil vapor then goes through a packed column, which is loaded with catalyst. This takes place in a catalytic cracking reactor within a temperature range of 300 to 500°C. Post the cracking reaction is complete, a steel separator is used to bifurcate the bio-oil and condensed gas.

Molecular sieves such as ZSM-5 (Zeolite SoconyMobil-5) is used commonly in catalytic cracking processes. A molecular sieve is a tool which has minute holes having precise and uniform dimensions. These holes allow small molecules to pass while blocking large molecules.

The basic structure of molecular sieves has a three-dimensional interconnecting network of alumina tetrahedra and are made of crystalline metal aluminosilicates. Since alumina is not present, molecular sieves can provide acidic sites when the cracking reaction takes place. This leads to the structural arrangement of the reactants.

## 8.4.3. Solvent Liquefaction of Lignocellulosic Biomass

1.    Current Liquefaction Process

Biomass liquefaction refer to the decomposition process of lignocellulosic materials which takes place in organic/aq solutions under pressure. During liquefaction, several solvents such as phenol, $H_2O$, alcohol, and multiple hydroxyl compounds are effective as reaction media. The liquid products which are derived possess better fuel properties and higher conversion rates than the bio-oil derived from fast pyrolysis reaction. They have lower moisture content, free from acids and higher heat value.

It seems that only those products liquefied in $H_2O$ /low carbon alcohols are proposed as suitable for biofuel production. Products of liquefaction in phenol and glycerol are mainly used as substitute materials, such as phenolic resins and polyurethane foams.

$H_2O$ is the cheapest solvent used in liquefaction. However, yields of the product usually are not satisfactory and several reports indicate that only 5%–40% of biomass conversion is achieved. Also, there is the need for treatment of $H_2O$-soluble compounds in order to avoid the deleterious environmental impact. Low-carbon alcohol is a potential solvent for liquefaction.

These alcohols not only can improve the yield, but also can be easily recovered by distillation after the reaction. Although the properties of liquefied oils are improved to a certain extent compared with fast pyrolysis oil, the properties are still not equivalent to those of petroleum fuels.

Additional upgrading processes, such as hydro processing, are needed, which yields a product very similar to the bio-oil generated from fast pyrolysis reaction.

2.    Integrated Liquefaction

This production of phenolic compounds can be done through liquefaction of biomass. During the process, biomass can also yield monosaccharide derivatives. This is because, biomass is composed of cellulose, hemicellulose, and lignin. The decomposition products obtained from lignin are distinct from cellulose and hemicellulose.

The liquefied products which are derived from cellulose and hemicellulose offer sufficient hydroxyl functionalities. This is due to the fact that the original units are comprised of a large number of hydrophilic groups some of which are carbonyl groups and hydroxyl groups.

On the other hand, decomposition products derived from lignin are generally made of phenol derivates. An efficient method has not been developed yet to segregate these two groups of liquefied products. Many oxygenated compounds are present in the liquefied products, as a result of which the cost of separation methods used like distillation and extraction increases. The yields reduce as a result and the significance of co-products utilization and disposal increases.

A Chinese Academy for Forestry has proposed an alternative solution which involves integrated liquefaction of biomass. Microwave energy is used along with acidic catalyst which yields only two groups of liquefied products, sugar derivatives and phenolic compounds having similar molecular structure.

A fractionation method was proposed for sugar derivatives and phenolic compounds with a purity of 80% and 65% for sugar derivatives and phenolic, respectively which aimed to produce two kinds of products close to their original molecular structure. This is a simple approach which can enable an integrated use of liquefied products, on the basis of their molecular structures, toward different synthesis directions.

## 8.4.4. Future Directions

Biomass is the only renewable source of carbon from which liquid fuels can be obtained through thermochemical conversion. However, to do this on a commercial scale requires availability of feedstock and catalysis system and catalytic reactors. Though many studies have been conducted, but challenges are present.

There technological barrier with respect to the conversion of lignocellulosic biomass in scaled up process which is due to the complex nature of the cellulose and lignin in the biomass. As a result, during thermochemical conversion reaction pathways are unpredictable. Hence a large amount of undesirable product is generated which are detrimental to the fuel properties of pyrolytic oils.

There is ambiguity regarding the mechanisms which also present a challenge for scaling up facilities to enable fast pyrolysis of lignocellulosic biomass. The feeder system and the condenser system get clogged by the formation of tar, and hence not suitable for long-term operation.

These challenges can be overcome in two ways. The first is by developing a an extremely efficient catalyst system which is capable of supporting multiple reactions using various reactive groups such as carboxylic group, aldehydes, and ketones, in liquefied biomass. This can generate molecular structures with the desired fuel properties, such as esters, alkanes, and alkenes.

The quality of the bio-oils is poor to improve which, reduction in the cost of catalyst is needed in terms of the activity, selectivity, and development of micro/mesoporous composite catalysts, and silica alumina composite catalysts which have along life and are stable. Another technique is to produce predictable products during the liquefaction process. Directional liquefaction could pave the way for production of high-grade liquid biofuels.

## 8.5. THERMOCHEMICAL WATER SPLITTING FOR HYDROGEN PRODUCTION

Thermochemical $H_2O$ splitting is performed in high temperatures and chemical reactions. Heat is generated from concentrated solar power or power generated from waste heat of nuclear power reactions. This is a long-term technology pathway which does not involve greenhouse gas (GHG) emissions.

The increasing concerns with respect to global climate change and the scarcity of non-renewable fossil fuels is leading to an increased demand for renewable sources. Studies reveal that the solar energy that the Earth absorbs for an hour can satisfy the energy requirements for a year. However, hydrogen is also emerging as a viable alternative fossil fuel because of its high energy density and non-polluting end products.

Hence, the technique of using solar power to split $H_2O$ into $H_2$ and synfuel using $H_2O$ and $CO_2$ is a viable option in this context, both these compounds being renewable resources.

Approaches like artificial photosynthesis and photovoltaic-powered electrolysis of $H_2O$ are also being developed. However, they are not widely implemented due to the low solar-to fuel conversion efficiency (solar-to-fuel) of) 5% and <15% respectively.

Solar thermochemical process is seen an efficient process through which large scale production of $H_2$ can be done by making use of an entire solar spectrum. In 1980s, studies started being conducted in thermochemical splitting of $H_2O$ and several thermochemical cycles started being examined.

The thermochemical methods can be classified under two main categories, the low-temperature multistep processes and the high-temperature two-step processes.

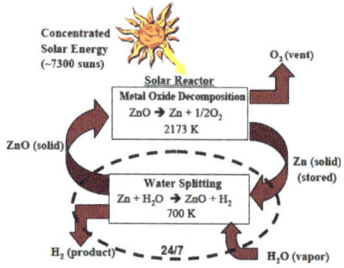

**Figure 8.8.** Thermochemical water splitting releases hydrogen.

*Source: Image by Wikimedia Commons.*

## 8.5.1. Low-Temperature Multistep Cycles

Low-temperature cycles have less radiative loses. Moreover, there are several heat resources that can be used to fuel the process. Several studies have been carried out in the past decade in this area such as the performance of cycles like S–I and S–Br cycles as well as Fe–Cl, Hg–Br, and Cu–Cl cycles. Though they can be used for steady production of H2 there are some environmental issues associated with the process like separation of acid mixtures, heavy-metal processing, decomposition of acids, generation of toxic or corrosive by-products and many others.

### 8.5.1.1. Manganese Oxide-Based Cycles

Transition metal oxide-based multistep cycles have become popular. One such cycle is the four-step cycle based on manganese oxides. In this cycle, Mn2O3 reduces to MnO and O2 is produced above 1,500°C. In the next step NaOH oxidizes Mn(II) back to Mn(III) as NaMnO2(s) and H2 is produced above 600°C.

The best efficiency of this cycle was 74% when 100% heat recovery is there and 16–22% when 0% heat recovery is there. NaOH is volatile above 800°C and high conversion temperature is required for Mn2O3-to-MnO conversion which makes implementation of this cycle difficult.

Tamaura et al. 1998 suggested the use of Na2CO3 for producing H2 from redox active MnFe2O4. Though this cycle closes below 1,000°C, the stoichiometric quantity of O2 is not evolved. This is because Na+ is not completely extracted even when CO2 is present.

Mn3O4-Na2CO3-MnO–Based Cycle

The thermochemical cycle suggested by Davis and his co-workers is noteworthy here. It is based on Mn3O4/MnO oxides. The four reactions involved in the cycle is shown in the diagram below.

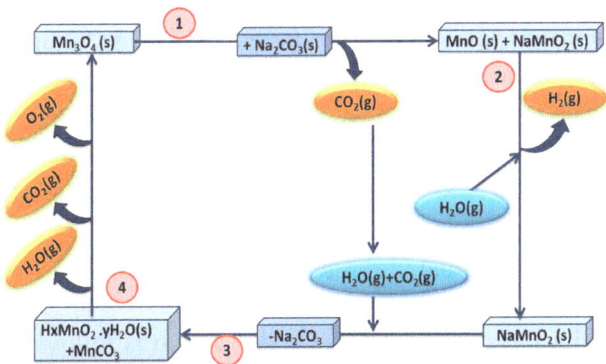

**Figure 8.9.** The Mn3O4-Na2CO3-MnO–Based Cycle

H2 is produced at a temperature of 850°C in this cycle and does not involve any corrosive products. Use of nanoparticles of Mn3O4 and Na2CO3 can help in drastically reducing the CO2 evolution. The slow evolution of H2 is the main hurdle in this cycle which is caused by the use of nanoparticles which can bring down the temperature to 750°C or 700°C, although the H2 evolution rate is slower in comparison with that at 850°C.

Na+ is extracted from layered NaMnO2 using hydrolysis under a CO2 atmosphere at 80°C for 3 h. Birnessite is the by-product and it reduces to Mn3O4 with stoichiometric O2 evolution.

## 8.5.2. Two-Step Thermochemical Processes

The two-step metal oxide process uses solar concentrators. The H2 and O2 need not be separated in this process. The metal oxide reduces the metal to a lower valent metal oxide. O2 is released during the endothermic step. In the subsequent step, it reacts with H2O and is re-oxidized. In the process a stoichiometric amount of H2 is released.

The thermal reduction of oxide and splitting of H2O have been performed at the same temperature where Tred = Toxd = Tiso. There is a large pressure swing in the gas composition between oxidation and reduction at this temperature which facilitates the process.

The process of splitting CO2 can be compared to the process of splitting of H2O in which a stoichiometric quantity of CO is released. Hence this is process is useful producing syngas(H2+CO+CO2). This can be converted and used as liquid fuel.

Splitting CO2 also reduces the concentration levels of the gas. CO2 splitting is thermodynamically more favoured than H2O splitting. It also gents the kinetic advantages of reduction. The two reaction, H2O –gas shift reaction and the reverse H2O –gas shift reaction has emerged as alternate techniques to produce liquid fuel using the production of syngas. The two reactions are H2O +CO $\rightarrow$ CO2 + H2 and H2 +CO2 $\rightarrow$ CO + H2O, respectively.

Maravelias et. al evaluated three main thermochemical fuel production systems. In each of the system, syngas is produced by a different method. In the first one solar CO2 splitting takes place which is followed by the H2O -gas shift reaction. In the second one, solar H2O splitting takes place and then reverse H2O –gas shift reaction occurs. In the third method. splitting of CO2 and H2O occurs without either of the reaction.

In the first two methods, the systems are less efficient because the CO2/CO separation is an energy intensive process. Moreover, the thermodynamics of the reverse H2O gas shift reaction is unfavourable. The efficiency of system in the third method is high, because there less requirement for CO2/CO separation. Moreover, the H2:CO ratio is set by the distribution of the solar heating units.

Both stoichiometric and nonstoichiometric oxides can be used to make two-step cycles. The former may be volatile of non-volatile when redox takes place. The volatile cycles in most cases are ZnO/Zn, SnO/SnO2, In2O3/In, and CdO/Cd redox pairs. During the cycling, they exhibit reversible solid–gas phase transitions. In volatile cycles, high entropic gains occur when gaseous product are formed during reduction which makes it thermodynamically favourable.

On main challenge, which has to be overcome in these cycles is the recombination of product gases. In non-volatile cycles the redox pairs do not encounter the problems of product combination since they remain in the condensed state.

The Fe3O4/ FeO which was proposed by Nakamura, 1997 needs a temperature of 2,500 K. The melting point of these two compounds are 1,870 and 1,650 respectively which is lower than its reduction temperature. As a result, severe coarsening of particles occurs which deactivates the cycle further.

Divalent metal ions such as Zn, Ni, Co, Mg, and Mn were incorporated in the $Fe_3O_4$ matrix which led to the formation of mixed ferrites and lowered the reduction of temperature.When Co and Ni are used, the reduction capability of ferrites improves. When NiO is used in the $Fe_3O_4$ matrix, the performance improves. The redox activity of ferrites is retarded due to slow diffusion of $Fe^{2+}$. This limitation can be overcome either by nano structuring of iron oxides or introducing additional support.Nano structing has been helpful in improving the reaction kinetics. This occurs due to the deposition of atomic layer of $CoFe_2O_4$ and $NiFe_2O_4$ on high surface area $Al_2O_3$. However, this is not a practical solution, since coarsening of ferrite particles happen.

### 8.5.3. Future Outlook

Solar thermochemical technique for generating hydrogen is a viable process. However, there are several challenges in this area, which makes it important to carry out a research on a large-scale for generating hydrogen using low-temperature, multistep thermochemical cycle based on Mn (III)/Mn (II) oxides. A pilot project can be set up for splitting $CO_2$ and $H_2O$.

The by-product syngas helps to produce organics. Thermodynamic analysis of various oxides can help in preliminary materials screening. Kinetics is an important consideration for calculating efficiency. The reactant gases should not be used much, or the process of their separation should be more energy efficient. Other ways of improving the technique are by designing solar reactors and using porous material to address the challenge of heat transfer-limited reduction.

Extensive research in this technology can help production of hydrogen on a large scale through splitting $H_2O$ a viable option. This will also address the important concern of changing climate.

## 8.6. THERMOCHEMICAL ENERGY STORAGE TECH-NOLOGY

The problem of excessive fuel consumption of fossil fuels and the scarcity of non-renewable sources is becoming severe and has an impact on sustainable development. New methods of thermal energy storage (TES) are being developed with the aim of harvesting renewable energy sources.

The main challenge in this respect is to effectively store excess energy and bridge the gap between energy generation and consumption. Thermochemical energy system is a promising technology which can help achieving this objective by reducing the time or rate mismatch between the energy demand and supply. It is a high ES density technology and hence is emerging as a popular technique.

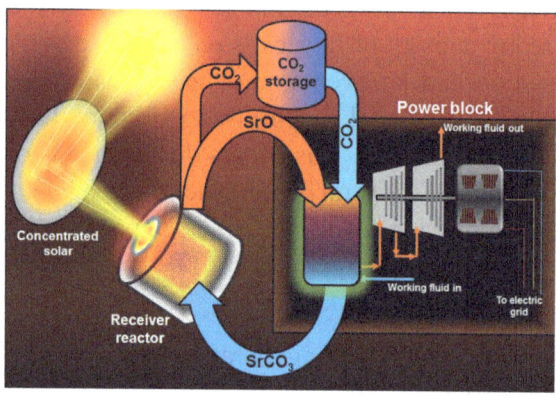

**Figure 8.10.** Thermochemical energy is a viable option for bridging the gap between demand and supply of fuel.

*Source: Image by Flickr.*

The Swedish and Swiss researchers pioneered the study of thermochemical energy storage using chemical heat pump. The first practical example of thermochemical heat storage was demonstrated in Germany in 1980s. During the period 1842–92, a project on super heat pump and energy-integrated system was conducted in Japan.

In 1991, CaO/H2O chemical heat pump was used to examine an enhanced thermal heat transfer. Several reactant mixtures were examined in Japan, France, and UK to enhance the thermal heat transfer of adsorbent bed.

## 8.6.1. The Need for Thermochemical Energy Storage

At present, demands of societal energy are increasing whereas fossil fuel resources, which dominate most national energy systems, are limited. In coming years, they are predicted to become scarcer as well as more expensive. Moreover, a number of concerns exist regarding the environmental impacts associated with increasing energy consumption. For example: atmospheric pollution and climate change.

The main cause of climate changes and agreements to limit them are considered as GHG emissions. For example: the Kyoto Protocol, have been developed. Partly through the adoption of advanced energy technologies and systems where advantageous, changes are required in energy systems to address serious environmental concerns.

The concern regarding environmental problems and anticipated worldwide increase in energy demand is fostering the utilization of more efficient as well as cleaner energy technologies in relevant applications. For instance: advanced systems for waste energy recovery and energy integration. TES has a widespread application and it is an important technology that can contribute to avoiding environmental challenges and increasing the efficiency of energy consumption.

The temporary holding of thermal energy in the form of hot or cold substances for later utilization is known as TES. As TES can make their operation more efficient, particularly by bridging the period between periods when energy is harvested and periods when it is needed, it is considered as a significant technology in systems involving renewable energies as well as other energy resources. In simple terms, TES is helpful for balancing between the demand as well as supply of energy.

Hence, we can say that TES plays a vital role in increasing the contribution of various types of renewable energy in the energy mix of regions as well as countries. A number of TES technologies as well as applications exist. There are many factors on which selection of a TES system for a particular application depends. The factors include storage duration, storage capacity, economics, supply, and utilization temperature requirements, heat losses and available space. Sensible and latent are the two main types of TES. Sensible TES systems store energy through altering the temperature of the storage medium.

It can be rock, $H_2O$, soil, brine, etc. Latent TES systems store energy by change of phase. For example, cold storage ice/$H_2O$ and heat storage through melting paraffin waxes. Sensible storage units are generally larger than Latent TES units. Based on storages that utilize chemical reactions, more compact TES can be achieved. Thermochemical storage systems which constitute the focus of this article have recently been the subject of increased attention.

Moreover, it could be especially advantageous where space is limited. However, thermochemical TES systems are not yet commercial. Thus, an appropriate research and development is needed to better understand as well as design these technologies and also to resolve other practical aspects before commercial implementation can take place.

A better understanding of their efficiencies is required in particular. Principles of thermochemical TES as well as recent advances are presented in this article. Thermochemical TES is critically assessed and also compared with other TES types. Moreover, merits, and demerits of thermochemical TES are considered as they are concerned with other TES types. To improve understanding and thereby support development and ultimate implementation of the thermochemical TES technology is its objective.

## 8.6.2. Type of Energy Storage

Energy storage is important for storing energy from any intermittent source to meet variable demand. It can be done using devices or physical media. Energy storage systems have complex properties which have to be evaluated properly. Advanced systems can be used to provide feasible innovation solutions by integrating it with other technologies. Energy storage devices can be classified as mechanical, thermal, chemical, biological, and magnetic.

## 8.6.3. Classification of Thermochemical Energy Storage

Advanced thermochemical energy storage has been successfully used in many countries, especially in developed countries. It is comprised of several technologies for storing heat and cold energy for utilizing them later. Thermal energy can be stored as latent heat, sensible heat, thermochemical heat and a combination of these materials. Some of the possible methods can be divided into physical and chemical process.

1.    Physical Processes

The physical energy storage process can be divided into sensible and latent energy storage. Sensible heat is thermodynamic process which can be calculated mathematically. It is the heat that is absorbed or released when a change in temperature of a substance takes place. In most cases, sensible heat systems make use of liquids like oil and H2O. In some cases, even rock, iron, molten salts and bricks are used, though not preferred due to their volumetric heat capacity of gases.

In latent heat storage, a chemical substance is used to release or absorb heat. This happens during change if phase from solid to liquid or gas to liquid, which is not accompanied by a change in temperature. This heat storage is more attractive than sensible heat storage. Since the latent heat change is higher than sensible heat change, therefore latent heat systems offer a significant storage density. They can store heat at a constant temperature which corresponds to the phase transition temperature of the heat storage medium.

2.    Chemical Processes

A chemical energy storage system has one or more chemical compounds. The latter includes sorption and thermochemical reactions. In thermochemical energy storage systems, a dissociation reaction is used to store energy and is recovered in a chemically reverse reaction. In these reactions the temperature of a few substances may increase or decrease. The chemical heat energy released in this manner can be stored for long term using effective methods.

The heat that can be stored is determined by factors like amount of storage material, the endothermic heat of reaction and the extent of conversion.

## 8.6.4. Principles of Thermochemical Energy Storage

The basis of both sorption and thermochemical storage systems are reversible chemical reactions. This is also known as desorption in which heat is absorbed during and endothermic decomposition process. In a reverse exothermic synthesis reaction, the stored heat is given back.

A thermochemical energy storage process generally includes three main processes:

Charging: In this endothermic reaction, an energy source is used for dissociation of a thermochemical material into two components which can be in any phase.

Storing: The products formed in the charging process, which are reactants as working pair or sorption couple, are stored separately in this stage

Discharging: This involves an exothermic reaction in which the two products are again combined, and the compound original thermochemical material is regenerated. The recovered energy is released in the process.

## 8.6.5. Chemical Storage and Sorption Storage Classification

Chemical TES can be classified into chemical storage and thermochemical storage. Thermochemical storage can be further divided into open and closed systems. There is no clear distinction between chemical storage, sorption storage and thermochemical storage. The diagram below shows the classification of chemical and sorption storage.

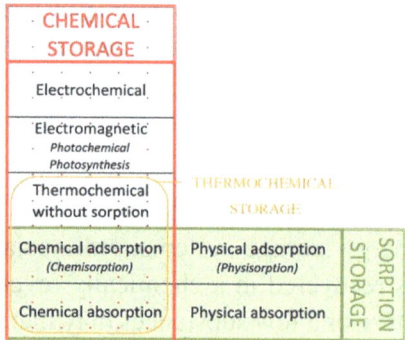

**Figure 8.11.** The relationship and distinction between chemical and thermochemical storage.

Chemical energy storage involves a chemical reaction through which chemical energy is transformed into other forms of energy. Sorption is a technique through which gas, or a vapor is captured or fixed by a substance called sorbent, which is generally in condensed state.

Sorption can be classified as physical or chemical sorption based on the type of bonding. It includes both absorption and adsorption, which are different phenomena. However, both of them involve the physical transfer of a volume of mass or energy.

Absorption is concerned with a transfer of a volume into a volume. In other words, it involves a permeation or dissolution of a volume of energy or mass into another volume of energy or mass. The former is called absorbate and the latter is called absorbent.

Open and Closed Sorption Systems

Sorption energy storage systems can be divided into open and closed systems.

## 1. Closed Thermochemical Energy System

The physical effect of closed sorption system is the same as open storage, but the engineering is quite different. The components of a closed storage cannot be exposed to atmosphere. The sorption process uses H2O vapor as adsorptive. Adjustments can be made to the operation pressure of the working fluid. A heat exchanger is used to transfer heat energy from the systems to and from the adsorbent. A closed system has a lower expected energy density.

Abedin and Rosen, 2011, assessed closed thermochemical energy system using energy and exergy. A closed storage system has to be charged using higher temperatures. Levy et al., 1993, investigated a combined solar energy storage system with a closed loop chemical heat pipe and found the overall performance of the closed loop satisfactory.

## 2. Open Thermochemical Energy System

An open system has a bed of solid adsorbents or a reactor where the air is exposed to a liquid desiccant. Air carries the H2O vapor and heat energy in and out of this packed bed. The system has a working fluid and thermochemical material. The gaseous working fluid is released into the atmosphere where it operates at the atmospheric pressure.

H2O is most suitable as working fluid. The materials used should also be non-flammable and non-toxic. These systems use separate desorption step and adsorption step to store thermal energy. This eliminates the possibility of thermal energy loss.

Abedin and Rosen, 2011investigated both these systems. When SrBr·6H2O were used as thermal materials the overall energy and exergy efficiency were 50% and 90% respectively. When Zeolite 13X was used as thermal material, the overall energy and exergy efficiencies are 69 and 23%, respectively.

The exergy method enhances assessment of energy method. This provides a significant margin for improving efficiency and reducing loss for open storage systems.

Wu et al, 2009, evaluated, and numerically analysed an open thermal storage system with the help of composite sorbents. Experimental measurements on open thermochemical energy system was conducted to validate the computation results. The specific TES capacity increased noticeably while the coefficient of performance of the system decreased.

## 3. Chemical Heat Pump Storage Systems

Chemical heat pump is mainly concerned with chemical thermal energy conversion and storage. It uses the transformation that takes place between thermal power and potential energy. Specifically, chemical heat pumps change the temperature of the thermal energy which are stored by chemical compounds by using the reversible chemical reaction and sorption.

These chemical materials are primarily responsible for absorbing and releasing heat. The nature of the chemical reaction determines the type of chemical materials that can be used. A chemical heat pump system can be both mono-variant system and di-variant.

The chemical heat pumps can be classified on the basis of phase of working pair into solid to gas and liquid to gas. The former system has reactor, evaporator, and condenser. The latter is made of at least two reactors, a condenser and evaporator.

Several researches have been undertaken in the past decades, which investigated the design and fundamental components and process optimization, the appropriate materials, the behaviour and field performance of these thermochemical energy systems, both in the long and short term.

Currently, some challenges and barriers restrict the usage of these systems on a large scale. Chemical energy storage is at its nascent stage. The technology is yet to be applied into practice though many patents have been filed. There are several technical and economical questions, which require further investigations. The long-term efficiency and viability of the thermochemical energy systems can be proved through further research and development and implementation of large-scale demonstration project.

## 8.6.6. Advantages of Thermochemical Energy Storage

Over other types of TES, thermochemical TES systems have several advantages:

- After cooling to ambient conditions subsequent to their formation, components (A and B) can usually be stored separately at ambient temperature. Hence, there is little or no heat loss during the storing period. As a result, there is no need of insulation.

- Thermochemical TES systems are especially suitable for long-term energy storage (e.g., seasonal storage), as a consequence of the low heat losses.

- Thermochemical materials possess higher energy densities as compared to PCMs as well as sensible storage media. Thermochemical TES systems can provide more compact energy storage as compared to latent and sensible TES because of higher energy density. This feature proves useful where space for the TES is valuable or limited.

## 8.7. CONCLUSION

Thermochemistry is concerned with the energy changes that takes places when chemical reactions occur. Thermochemistry has applications in various areas. The isothermal titration calorimeter works on the principles of thermochemistry. The enthalpy diagrams are widely used compare groups of substances which have some common features. It also helps in calculating the energy content of fuels. Thermochemistry has its usage in the field of nutrition where it helps in calorific value of food substances based on its composition.

The fundamentals of thermochemistry provide the answer to many pressing concerns like the scarcity of fossil fuels, emission of GHGs during certain processes, addressing the energy storage issues. Various technologies have emerged like thermochemical conversion of lignocellulosic biomass for obtaining fuel, splitting of $H_2O$ for producing hydrogen and energy storage technology. However, these technologies are still in their nascent stage and require extensive research and large-scale pilot project for evaluating its efficiency.

# REFERENCES

1.  Abanades, S., (2019). Metal oxides applied to thermochemical water-splitting for hydrogen production using concentrated solar energy. Chem. Engineering, 3(3), 63.

2.  Auyeung, N., & Kreider, P., (2017). Solar Thermochemical Energy Storage. [online] Aiche.org. Available at: https://www.aiche.org/resources/publications/cep/2017/july/solar-thermochemical-energy-storage (accessed on 13 March 2020).

3.  Energy.gov. (n.d.). Hydrogen Production: Thermochemical Water Splitting. [online] Available at: https://www.energy.gov/eere/fuelcells/hydrogen-production-thermochemical-water-splitting (accessed on 13 March 2020).

4.  Jiang, J., Xu, J., & Song, Z., (2015). Review of the direct thermochemical conversion of lignocellulosic biomass for liquid fuels. Frontiers of Agricultural Science and Engineering, 2(1), 13.

5.  Lower, S., (2019). 14.6: Applications of Thermochemistry. [online] Chemistry LibreTexts. Available at: https://chem.libretexts.org/Bookshelves/General_Chemistry/Book%3A_Chem1_(Lower)/14%3A_Thermochemistry/14.06%3A_Applications_of_Thermochemistry (accessed on 13 March 2020).

6.  Rao, C., & Dey, S., (2017). Solar thermochemical splitting of water to generate hydrogen. Proceedings of the National Academy of Sciences, 114(51), 13385–13393.

7.  Zsembinszki, G., Solé, A., Barreneche, C., Prieto, C., Fernández, A., & Cabeza, L., (2018). Review of reactors with potential use in thermochemical energy storage in concentrated solar power plants. Energies, 11(9), 2358.

# INDEX

Printed in the United States
By Bookmasters